INDIA'S
HABITUATION
WITH THE
BOMB

Nuclear Learning in South Asia

INDIA'S
HABITUATION
WITH THE
BOMB

Nuclear Learning in South Asia

Edited by
NAEEM SALIK

OXFORD
UNIVERSITY PRESS

OXFORD
UNIVERSITY PRESS

Oxford University Press is a department of the University of Oxford.
It furthers the University's objective of excellence in research, scholarship,
and education by publishing worldwide. Oxford is a registered trade mark of
Oxford University Press in the UK and in certain other countries

Published in Pakistan by
Oxford University Press
No. 38, Sector 15, Korangi Industrial Area,
PO Box 8214, Karachi-74900, Pakistan

© Oxford University Press 2019

The moral rights of the author have been asserted

First Edition published in 2019

ISBN 978-0-19-070139-0

Typeset in Adobe Garamond Pro
Printed on 55gsm Book Paper

Printed by Delta Dot Technologies (Pvt.) Ltd., Karachi

Contents

Foreword   vii
Acknowledgements   x
List of Acronyms   xi

Introduction   1

1. A Brief Historical Overview of India's Nuclear Programme   18
   Naeem Salik

2. India's Nuclear Doctrine: Stasis or Dynamism?   59
   Ali Ahmed

3. Nuclear Weapons Governance in India: How Robust   89
   and Stringent?
   Sitakanta Mishra

4. India's Export Control Regime   109
   Zafar Ali

5. India's Nuclear Regulatory Regime   130
   Happymon Jacob and *Tanvi Kulkarni*

6. India's Ballistic Missile Defence Programme: Implications 167
   for Strategic Stability
   Naeem Salik

Conclusion   186

Contributors   192
Index   197

Foreword

It gives me great pleasure to write a foreword for Dr Naeem Salik's book on a significant and absorbing subject, 'nuclear learning', especially in South Asia. As former Director General, Strategic Plans Division (SPD), I had been personally involved with the development, consolidation, and maturing of Pakistan's nuclear programme. I am happy to say that at the SPD, every day one learnt something new, as we moved speedily towards the consolidation and strengthening of the nation's nuclear capability in a variety of dimensions against an evolving threat in the post-May 1998 era. I had the privilege of learning much from my personal association with Dr Naeem Salik when we served together in the newly established SPD in late 1998; later, Dr Salik assumed the appointment of Director, Arms Control and Disarmament Affairs (ACDA).

For all of us who came from mainstream soldiering to the business of nuclear management, the subject of arms control and disarmament—with implied linkages to international nuclear diplomacy—was not only quite complex and fascinating, it also required an in-depth understanding and background of various ACDA issues. Undoubtedly, there was no one better equipped in terms of background knowledge and understanding of the nuances of the subject than Dr Naeem Salik. In many ways, therefore, it is a good opportunity for me to acknowledge and put on public record not only Dr Salik's outstanding services to Pakistan's nuclear programme but also my gratitude for his contribution in my personal learning and education in the area of ACDA. He left us in search of 'greater nuclear learning' and we were all that much poorer to lose his expertise and work ethos. But I am also glad that in Pakistan's larger interest, he pursued his subject with a single-minded focus, to emerge as an intellectual of national and international standing. Today, no worthwhile seminar or discourse on strategic thought is complete without his intellectual input. Thank you, Naeem!

The concept of nuclear learning is not widely understood amongst researchers and scholars, particularly in South Asia. Few writers have written about nuclear learning and fewer still have published well-researched articles on this phenomenon in South Asia.

Cognitive learning theory is a process of acquisition, absorption, and retention of information. Learning takes place through internalising information and the level of learning is assessed from the changes in the behaviour of an individual or an organisation involved in the learning process. Learning, therefore, acts as the stimulus that brings about change in behaviour. Interestingly, the manifold implications of states possessing nuclear weapons act as a powerful stimulus that shapes the behaviour of national institutions dealing with nuclear issues and the political discourse in a country. I have always believed that there are no upper limits to learning and education and therefore there is a need to keep minds open and flexible to new ideas and developments to retain dynamism in the organisation and in the leader's thoughts.

The concept of nuclear learning entails a series of steps that nuclear-armed states undertake comprehensively for the effective development and management of their nuclear capability. The measures taken by nuclear-capable states can range from nuclear-specific laws; command and control systems; physical protection of nuclear material and their delivery systems; doctrinal, operational, and force development issues; and not the least, at the overarching foreign policy levels, the issues of arms control and disarmament as they might affect national nuclear policies and programmes.

India and Pakistan both became declared nuclear weapon states in May 1998. Since then, because of my vantage point, I am confident that Pakistan has taken sound measures to manage its nuclear assets better in a variety of dimensions. Pakistan's efforts continue to be acknowledged by responsible elements in the international community; however, one would very much like to believe that like Pakistan, India too has taken measures to better manage its nuclear assets.

Dr Salik's well-researched book, *Learning to Live with the Bomb—Pakistan: 1998–2016* (OUP, 2017) gave an accurate and detailed description of Pakistan's nuclear learning—focused mainly on the positive learning curve within the SPD and the state at large—of over nearly two decades. Since its publication, however, many amongst the nuclear analysts' community felt the absence of a corresponding book on India's nuclear learning. The present volume, *India's Habituation with the Bomb: Nuclear Learning in South Asia*, fills the void to a certain extent. Edited by Dr Naeem Salik, it contains chapters contributed by four Indian (Ali Ahmed, Sitakanta Mishra, Happymon Jacob, and Tanvi Kulkarni) and two Pakistani (Naeem Salik and Zafar Ali) scholars. Not a typical book on the history of nuclear weapons development by the two South Asian adversaries, it provides an analysis of both states' journey to the status of responsible nuclear powers. Additionally, a concluding chapter by Dr Salik gives the readers a balanced and objective analysis of nuclear learning in India and Pakistan. By putting together the work of five authors in one place, Dr Salik has balanced his earlier book. I would like to hope that in the coming years, more scholars would be encouraged to write about this particular subject to further enrich literature on nuclear learning in South Asia, which remains an ongoing process and a work in progress.

Lieutenant General (retd) Khalid Ahmed Kidwai NI, HI, HI (M)
Advisor, National Command Authority
Former Director General, Strategic Plans Division

Acknowledgements

The completion of this study would not have been possible without the unstinted moral, administrative, and financial support provided by Ambassador (retd) Ali Sarwar Naqvi, Executive Director, Center for International Strategic Studies (CISS), Islamabad. I would also like to acknowledge the invaluable contributions of my co-authors. Without their diligence and cooperation, this project could not have been completed. I also owe a debt of gratitude to my colleagues at CISS for their continued encouragement and my family for allowing me a major share of their time to concentrate on my work.

List of Acronyms

AAD	Advanced Air Defence/ Ashwin Ballistic Missile Interceptor
ABM	Anti-Ballistic Missile
AEC	Atomic Energy Commission
AERB	Atomic Energy Regulatory Board
AG	Australia Group
ANPR	American National Public Radio
BARC	Bhaba Atomic Research Center
BDB	Beyond Design Basis
BJP	Bharatiya Janata Party
BMD	Ballistic Missile Defence
C&C	Command and Control
C3I	Command and Control, Communications, and Intelligence
CBM	Confidence Building Measure
CCS	Cabinet Committee on Security
CMD	Cruise Missile Defence
CNS	Convention on Nuclear Safety
COSC	Chief of Staff Committee
CPPNM	Convention on the Physical Protection of Nuclear Material
CTBT	Comprehensive Test Ban Treaty
DAE	Department of Atomic Energy
DGFT	Director General Foreign Trade
DIA	Defence Intelligence Agency
DRDO	Defence Research and Development Organisation
ENDC	Eighteen Nation Committee on Disarmament
ENR	Enrichment and Reprocessing

ENSREG   European Nuclear Safety Regulators Group
ERC   Event Review Committee
ERC   Emergency Response Centre
ERT   Emergency Response Team
FBR   Fast Breeder Reactor
FTDR   Foreign Trade, Development, and Regulation
GNSR   Global Nuclear Safety Regime
IAEA   International Atomic Energy Agency
IAMD   Integrated Air and Missile Defence
ICBM   Intercontinental Ballistic Missile
ICG   Intelligence Coordination Group
IECR   International Export Control Regimes
IGMDP   Integrated Guided Missile Development Program
INRA   International Nuclear Regulators' Association
INSAF   International Network for Safety Assurance of Fuel
               Cycle Industries
IPFM   International Panel on Fissile Materials
IRRS   Integrated Regulatory Review Service
J&K   Jammu and Kashmir
KARP   Kalpakkam Atomic Reprocessing Plant
LoC   Line of Control
LIS   Laser Isotope Separation
MAD   Mutually Assured Destruction
MEA   Ministry of External Affairs
MECR   Multilateral Export Control Regime
MOX   Mixed Oxide Fuel
MTCR   Missile Technology Control Regime
NAPS   Narora Atomic Power Station
NDA   National Democratic Alliance
NDRF   National Disaster Response Force
NEA   Nuclear Energy Agency
NFU   No First Use
NPP   Nuclear Power Plant
NSA   National Security Advisor

NSAB	National Security Advisory Board
NSC	National Security Council
NSCS	National Security Council Secretariat
NSG	Nuclear Suppliers Group
NSRA	Nuclear Safety Regulatory Authority
NSSP	Next Steps in Strategic Cooperation
PAC	Public Accounts Committee
PAD	Prithvi Air Defence
PAL	Permissive Action Link
PDV	Prithvi Defence Vehicle
PHWR	Pressurised Heavy-Water Reactor
PNE	Peaceful Nuclear Explosion
PTBT	Partial Test Ban Treaty
R&AW	Research and Analysis Wing
RTI	Right to Information
SALT	Strategic Arms Limitations Talks
SASA	Strategic Armament Safety Authority
SCOMET	Special Chemicals, Organisms, Materials, Equipment, and Technologies
SEC	Strategic Export Control
SFC	Strategic Force Command
SLBM	Submarine-Launched Ballistic Missile
SPD	Strategic Plans Division
SPS	Strategic Planning Staff
SSBN	Ship Submersible Ballistic, Nuclear
STC	Strategic Trade Control
TCG	Technical Coordination Group
TNWs	Tactical Nuclear Weapons
UA	United Alliance
WA	Wassenaar Arrangement
WANO	World Association of Nuclear Operators
WENRA	Western European Nuclear Regulators Association
WMD	Weapons of Mass Destruction

Introduction

India and Pakistan have projected themselves as responsible nuclear states in the international arena since their overt nuclearisation in May 1998. To give credence to their claims of responsible custodianship of their respective nuclear assets, they had to learn the intricacies of nuclear management, establish requisite institutional structures, and enhance their conceptual understanding of nuclear policies, doctrines, and postures. In India, despite the fact that its first nuclear test in May 1974 had been portrayed as a 'peaceful nuclear explosion' (PNE), some individuals amongst the civil and military bureaucracies as well as some academics had explored various scenarios where nuclear weapons could possibly be used in a conflict with either Pakistan or China. They analysed the implications of a nuclearised security environment for India and the constraints it would impose upon its use of conventional force, especially against Pakistan.

One of the strands of thought that emerged out of such discourse was the suggestion to adopt a 'No First Use' (NFU) policy mainly to gain political mileage. That idea was reflected in India's Draft Nuclear Doctrine, pronounced in August 1999, and in the later policy document of January 2003. However, despite the availability of this body of literature and people who had ventured into this domain, it took India over a year after its declaration of status as a nuclear weapons state to formulate its draft nuclear doctrine and another two and a half years to outline its nuclear command and control structure and finalise its nuclear doctrine. The recent debate in India, triggered by a statement by the Indian defence minister, Manohar Parrikar,[1] and a book written by former foreign secretary and National Security Advisor Shiv Shankar Menon,[2] indicates that India is moving up

1

along the nuclear learning curve, reviewing, and modifying its earlier precepts in light of the changing security landscape.

Since learning is a cognitive phenomenon, it is a challenging task to estimate the degree or extent of learning with a high degree of objectivity. However, it is possible to have a fair estimate of learning by evaluating the manifestations of learning in terms of the setting up of purpose-built institutions and the conduct of states by studying their crisis behaviour and evolution of security policies over a period of time. It is also difficult to separate learning from adaptation as they happen alternately and sometimes even simultaneously. Nuclear learning can be defined as under:

> Nuclear learning is a process through which states that acquire nuclear weapons capability learn to manage it through the development of nuclear doctrines, command and control structures, safety and security mechanisms, [and] regulatory regimes and acquire an understanding of both the technological characteristics of these weapons as well as their politico-strategic ramifications. This enables them to achieve a stable strategic balance through an astute manipulation of these formidable instruments of power.[3]

Nuclear learning itself is a subset of the broader concept of learning and various scholars have argued about the nature and levels of learning. For instance, Jeffery Knopf insists that learning must be 'normative' in nature to be considered as learning while others like Michael Levy argue in favour of a value neutral evaluation of learning. Nuclear learning takes place at various levels, starting with the individual, to institutional, to governmental, to state, and, finally, to international learning. Specifically in the nuclear domain, learning can also be categorised as 'factual learning', which means learning about the technical characteristics of nuclear weapons, and 'inferential learning' which implies an understanding of the broader policy implications of the acquisition of nuclear weapons. It can also be 'simple learning' wherein an adjustment is made to the means being employed for the achievement of national objectives or 'complex learning' that requires a re-evaluation and modification of the objectives themselves. The available literature suggests that in most cases the learning is simple learning rather than complex learning. While it is apparent that in South Asia the adversaries have yet to review and modify their respective objectives in view of the nuclearisation of the region, they seem to have made some alterations in the means they employ, signifying a degree of simple learning as is evident from their management of a series of post-nuclearisation crises.

Nuclear Learning in South Asia

India and Pakistan declared themselves as overt nuclear states after the nuclear tests in May 1998. Prior to that, both countries had been following policies based on opacity about their respective nuclear capabilities mainly to deflect international pressure and sanctions. Such an approach, however, meant that public discourse about critical aspects of nuclear policy such as nuclear doctrines, command and control structures, safety and security arrangements, and regulatory regimes, was severely constrained. Post-1998, however, both countries moved rather swiftly to bring the conceptual aspects at par with their technological advancements. India pronounced its Draft Nuclear Doctrine on 17 August 1999, while Pakistan formally announced the details of its nuclear command and control structure in February 2000. Both countries embraced the credible minimum deterrence posture at the outset, ostensibly to avoid a nuclear/missiles arms race and to alleviate the concerns of the international community in this regard. India also professed a no first use doctrine. However, with the passage of time and as a consequence of technological developments, the respective force postures and doctrinal conceptions are witnessing some changes.

Nuclear Learning in India

Despite the fact that there was no official discourse on nuclear policy in the public domain, some individuals in both civilian as well as military bureaucracies were exploring ideas and concepts within their respective domains especially after India's first nuclear test in May 1974. Most prominent among these was K. Subrahmanyam who, as long serving Director of the Institute of Defence Studies and Analysis (IDSA), started writing on nuclear policy-related issues and also spoke on the subject at public forums. Within the military, General K. Sundarji as Commandant of the College of Combat organised a series of war games and table-top exercises which included some retired military officers as well as civilian experts to brainstorm various scenarios related to warfare, wherein one or both antagonists were nuclear armed states. He also published the proceedings of such deliberations in the form of 'Combat Papers'[4] which not only explained basic conceptual issues but also various contingencies of nuclear interaction in a third world scenario as far back as 1981. General Sundarji also wrote a book titled *Blind Men of Hindoostan* in 1983, which painted various scenarios of nuclear conflict with Pakistan in a fictional style but leaving a persuasive and powerful message for the Indian political leadership to take a firm decision regarding operationalisation of India's nuclear capability.[5]

The influential role played by Sundarji and Subrahmanyam in the evolution of Indian thinking on nuclear strategy has been acknowledged in an article by K. Sundaram and M. V. Ramana, especially in the adoption of the NFU policy, stating that:

> Thinking about how the Indian forces were to use nuclear weapons dates back to well before the nuclear tests of 1998 and a policy of No First Use has long figured in Indian discussions and debates on nuclear weapons, as well as regional diplomacy. This prominence is due, in large part, to the fact that two of the most influential voices in promoting the acquisition of nuclear weapons by India—General K. Sundarji, who later went to become the country's Chief of the

Armed Forces, and K. Subrahmanyam, a civil servant who directed the Institute for Defence Studies and Analyses for nearly a decade and a half—were votaries of an NFU policy.[6]

George Perkovich in his seminal work titled *India's Nuclear Bomb: The Impact on Global Proliferation*[7] has also recognised the central role played by General Sundarji in the development of India's nuclear weapons posture by initiating a public discussion on the subject in his capacity as Commandant of the College of Combat as far back as the early 1980s. His main argument revolved around the fact that nuclear weapons could pose a serious threat to the concentrated armoured formations—a key element in Indian and Pakistani strategy—which necessitated India's acquisition of nuclear weapons to deter such a threat because alternative options such as greater dispersal and manoeuvrability of forces would require much greater expenditure.[8]

In an unpublished dissertation titled 'Strategy in the Age of Nuclear Deterrence and Its Application to Developing Countries', submitted to the Madras University in 1984, Sundarji built a number of scenarios where he pitched India both with or without nuclear weapons against its traditional adversaries—China and Pakistan— arriving at the conclusion that nine of these scenarios would entail a first nuclear use and therefore, India could adopt a policy of no first use, especially once it has a clear advantage over Pakistan in terms of nuclear arsenals. This declaratory NFU policy, he believed, would ease India's transition from ambiguity to an overt nuclear posture.[9]

Perkovich points out that General Sundarji and K. Subrahmanyam were both part of important teams constituted in the 1980s to study matters related to nuclear weapons and to furnish necessary counsel to the Prime Minister. The first such task force was established when Prime Minister Rajiv Gandhi asked for an appraisal of the 'costs of a nuclear deterrent' in 1985.[10] The second committee was apparently set up on the instructions of Prime Minister V. P. Singh in 1990 to 'formulate procedures for effective control of the nation's nuclear arsenal and other issues related to nuclearisation'. Since V. P. Singh's government did not last long, the report prepared by this committee was presented to its successor, the Narasimha Rao government.[11] According to Perkovich the first committee reportedly suggested that India should build a 'minimal deterrent force', directed by an NFU doctrine, and meant solely for a retaliatory strike in response to a nuclear attack on Indian soil.[12] These instances clearly indicate that India had carried out serious studies and preparatory work for a prospective nuclear force especially after the 1974 test and well before the declaration of its overt nuclear status in 1998.

At the same time, some analysts such as Brigadier V. K. Nair had also undertaken comprehensive and detailed studies in the early 1990s in which he discussed practical issues like appropriate nuclear strategy for India, suggested nuclear command and control structures, targeting philosophy, possible nuclear targets in both China and Pakistan, the desired effects, and number and type of warheads required for each target. He arrived at a figure of 132 warheads for India to engage the probable targets in China and Pakistan. He also considered the weather and wind patterns in South Asia during various parts of the year and their likely impact on the nuclear fall-out.[13]

Unlike Pakistan, India, therefore, had a head start when it came to understanding the dynamics of security in a nuclearised environment as it could rely on individuals like K. Subrahmanyam, Air Commodore (retd) Jasjit Singh, and a number of academics and security analysts such as Bharat Karnad who had been contemplating the eventualities where a nuclear armed India would have to deal with two nuclear neighbours, namely China and Pakistan. Subrahmanyam's knowledge and expertise proved to be invaluable as Chairman of the National Security Advisory Board (NSAB) that was tasked with formulating India's prospective nuclear doctrine. Besides Subrahmanyam, several academic and security analysts who had researched and written on the nuclear issue were also part of the NSAB when it was first constituted in the aftermath of the nuclear tests.

The two decades since the nuclearisation of South Asia have seen a complete transformation of the regional as well as international security landscape, especially since the cataclysmic events of 9/11 that hastened the process of nuclear learning in South Asia. Though India and Pakistan had the advantage of hindsight and could learn and adapt the lessons from the experiences of earlier nuclear powers to their own peculiar environment, they have not been able to avoid the trial and error process of learning. They have thus navigated their ways through a clash of arms in the Kargil region along the Line of Control separating Indian and Pakistani controlled parts of Jammu and Kashmir, the prolonged military standoff in 2001/02 in the aftermath of an attack on the Indian Parliament, and the Mumbai crisis of 2008. It may, therefore, be interesting to study India's nuclear learning experience over the past two decades or so especially in the backdrop of developments such as the shifting nuclear competition between India and Pakistan into the maritime domain and the vigorous debate going on in India with some influential former officials demanding a review and modification of the Indian Nuclear Doctrine. India's learning would naturally be more complex than Pakistani learning since it has to contend with two nuclear rivals in China and Pakistan. It is inevitable that habituation with the bomb for two decades and experience of dealing with its politico-strategic implications would result in some learning which can be termed as 'experiential learning or learning by doing'. 'Experiential learning' can be described as:

> The most common form of learning that can also be found in many other spheres of human activity and comes about with the experience of dealing with nuclear weapons over a period of time—a kind of learning on the job. The degree of learning will vary amongst individuals as well as institutions depending on the availability, or lack thereof of a conducive environment for learning. Also, not everyone is likely to learn the same lessons from a similar experience. This type of learning is progressive and lessons learnt at an earlier stage of learning

continue to be modified and improved upon with further experience, availability of new information, or changes in security environment or technological developments.[14]

The book titled *Learning to live with the Bomb—Pakistan: 1998–2016*[15] detailed Pakistan's nuclear learning experience. It was then considered imperative to undertake a similar study about India to provide a companion publication that could enable scholars and students of international relations and security studies to carry out a comparative analysis of the quality and amount of learning in the two countries post-1998. The present volume consisting of contributions by both Indian as well as Pakistani scholars/experts—to maintain balance and objectivity—comprises the following chapters:

Introduction
Naeem Salik

This chapter briefly introduces the concept of 'Nuclear Learning' and places India's nuclear learning experience in the broader framework of nuclear learning. It is aimed at providing a backdrop to India's evolving nuclear policy in recent times to facilitate the understanding of the origin of some semi-permanent strands in India's nuclear discourse which had started well before the overt declaration of its status as a nuclear weapons-capable state in May 1998. It will also help identify the significant changes that have taken place post-1998 due to changes in the geo-strategic environment as well as technological advancements in the Indian nuclear and missile programmes.

Chapter 1 | A Brief Historical Overview of India's Nuclear Programme
Naeem Salik

Providing a broad historical overview of the evolution of the Indian programme and its underlying motivations, this chapter gives an insight into the benign neglect on the part of the global powers and their failure in constraining India's nuclear ambitions despite the availability of enough evidence to that effect in deference to their own geo-strategic interests. It helps in understanding the continuity and change in India's nuclear objectives from the inception of its programme to the present. It also provides a survey of India's sprawling nuclear enterprise and sets the stage for further insights into various aspects of India's nuclear learning experience in a variety of fields ranging from doctrinal thinking to command and control, safety and security, export controls, and the nuclear regulatory regime as well as the strategic implications of India's ambitions in the ballistic missile defence domain.

Chapter 2 | India's Nuclear Doctrine: Stasis or Dynamism?
Ali Ahmed

Ali Ahmed provides an incisive analysis of India's evolving doctrinal thinking, highlighting the incongruities between the declaratory and operational doctrines. He argues that it is unfair to believe that despite substantive changes in the strategic environment, India's nuclear doctrine has remained frozen since the last official pronouncement in January 2003. He is of the firm view that at the operational level, the doctrine has moved on in keeping with changes in the security environment. However, these changes have not been clearly articulated in public discourse despite some statements and writings by former senior officials, which has resulted in an apparently dichotomous situation and an image of the prevalence of inertia in India's nuclear doctrinal thinking. He tries to identify and highlight the possible contours of the actual nuclear doctrinal framework and to carry out a critical appraisal of the conceptual changes that appear to have taken place since the initial articulation of the doctrine. He points out that loosely used terminology such as 'massive retaliation' and mixing up first use with first strike and similar other slips reflect poorly on the understanding and competence of those responsible for formulating India's nuclear doctrine and policy. Ali Ahmed has also highlighted the anomalies in the existing chain of nuclear command wherein the commander of India's Strategic Force Command—a serving three-star General—reports to the civilian bureaucrats rather than to the political leadership, which is not an appropriate arrangement.

Chapter 3 | Nuclear Weapons Governance in India: How Robust and Stringent?
Sitakanta Mishra

Dr Mishra acknowledges that there is inadequate information available in public about India's command and control and safety and security of its nuclear assets, and that both these areas are surrounded with ambiguity. Partly, it is due to India's no first use posture and partly due to the need to maintain secrecy to ensure security of its nuclear weapons. He points out that the ultimate nuclear authority lies with the Prime Minister of India and is exercised through the office of the National Security Advisor to the Chairman of the Chiefs of Staff Committee and the Commander Strategic Force Command. Dr Mishra provides an outline of Indian nuclear command structure and mentions the existence of an alternate National Command Centre. He also tries to sketch out safety and security arrangements likely to be in place since there has been no official elaboration of the contours of the nuclear safety and security regime in place in India.

Chapter 4 | India's Export Control Regime
Zafar Ali

The current Director General of the Strategic Export Control Division, Ministry of Foreign Affairs, Pakistan, Dr Zafar Ali, has carried out a detailed survey of the evolution of India's Nuclear Export Control Regime as well as the development of an associated legislative framework for its implementation and enforcement. He highlights the fact that until the early 1990s, India's nuclear regulatory regime was very rudimentary and it was only after 1992 that the regime started developing in a more structured way and later brought dual use items in its export controls net. India promulgated its strategic export control law, Weapons of Mass Destruction and their Delivery Systems (Prohibition of Unlawful Activities) Act 2005 towards the end of 2005.

Chapter 5 | India's Nuclear Regulatory Regime
Happymon Jacob and Tanvi Kulkarni

An exposé of India's nuclear regulatory regime, this chapter points out that India has an expansive civilian nuclear programme, which is rapidly growing since India obtained the Nuclear Suppliers Group waiver in 2008 and it will expand further in the near future, which can only be managed with an effective regulatory regime. They trace the historical evolution of the International Nuclear Regulatory Regime, briefly describe India's nuclear regulatory regime, and then try to situate it within the broader framework of the international nuclear regulatory structure. Happymon and Kulkarni also describe the legislative and administrative framework related to nuclear regulation and identify various governmental agencies and their respective roles and functions in the regulatory mechanism.

Chapter 6 | India's Ballistic Missile Defence Programme: Implications for Strategic Stability
Naeem Salik

This chapter traces the scope, architecture, and technological developments that have taken place so far in the domain of India's ambitious Ballistic Missile Development Programme. It also deals with the larger issue of the impact of India's eventual operational deployment of this system on strategic stability in the region. Despite obvious limitations of current capabilities of missile defence systems under development in India and elsewhere in the world, its psychological impact on the adversaries cannot be understated. The author has pointed out the countermeasures already being instituted by Pakistan; China is bound to react in a similar way since the system would at least theoretically impinge upon the credibility of their respective deterrents against India. Such an offence–defence competition is likely to trigger an undesirable arms race that would result in adversely affecting the strategic stability in the region.

Conclusion
Naeem Salik

Based on an analysis of the earlier chapters, the conclusion enables the readers to have a clear sense of the state of nuclear learning in India since 1998 given the developments in diverse areas such as nuclear doctrinal discourse, legislative developments, and establishment of institutional structures for the effective management of India's nuclear capability. It provides an insight into the strengths and weaknesses of the existing systems in place and the successes and shortcomings in India's nuclear learning process. This in turn would provide, in future, a framework for a comparative analysis of respective nuclear learning processes in India and Pakistan in relation to the similar developments in other nuclear weapons states.

Notes

1. Special Correspondent, 'Why Bind Ourselves to "No First Use Policy", says Parrikar on India's Nuke Doctrine', *The Hindu*, New Delhi, 10 November 2016.

2. Shiv Shankar Menon, *Choices: Inside the Making of India's Foreign Policy*, Series: Geopolitics in the 21st Century (Washington DC: Brookings Institution Press, 2016).

3. Naeem Salik, *Learning to Live With the Bomb—Pakistan: 1998–2016* (Karachi: Oxford University Press, 2017), p. 5.

4. K. Sundarji, 'Effects of Nuclear Asymmetry on Conventional Deterrence', *Combat Papers I* (Mhow, MP, India: College of Combat, 1981); 'Nuclear Weapons in the Third World Context', *Combat Papers II* (Mhow, MP, India: College of Combat, August 1981).

5. K. Sundarji, *Blind Men of Hindoostan: Indo-Pak Nuclear War* (New Delhi: UBS Publishers Distributors Ltd, 1993).

6. K. Sundaram and M. V. Ramana, 'India and the Policy of No First Use of Nuclear Weapons', *Journal for Peace and Nuclear Disarmament*,

vol. 1, no. 1 (2018), pp. 152–68, <https://doi.org/10.1080/2575165 4.2018.1438737>

7. George Perkovich, *India's Nuclear Bomb: The Impact on Global Proliferation* (Berkeley, Los Angeles: University of California Press, 2001).

8. Perkovich quoted in Sundaram and Ramana, p. 153.

9. Sundaram and Ramana, pp. 153–4.

10. Perkovich quoted in Sundaram and Ramana, p. 154.

11. Gurmeet Kanwal quoted in Sundaram and Ramana, op. cit., p. 154.

12. Perkovich quoted in Sundaram and Ramana, p. 154.

13. Brigadier V. K. Nair, *Nuclear India* (New Delhi: Lancer Publishers, 1992).

14. Salik, op. cit., p. 14.

15. Salik, op. cit.

Annexure A

Nuclear Learning Typology*

Type of Learning	Illustration
Perceptual Learning	This kind of learning relates to the development of doctrinal precepts for the operationalisation of nuclear weapons and to integrate the conventional and nuclear war fighting concepts and doctrines.
Crisis Learning	This is another common form of learning and usually leads to substantive changes in the application of security policies. During the Cold War, the Cuban Missiles Crisis of October 1962 led to the adoption of various confidence building and crisis management mechanisms such as the Washington–Moscow Hotline, and created an environment for the beginning of arms control negotiations. In South Asia, the Kargil crisis of 1999 led to a better understanding of the complexities of a nuclearised security environment and had a salutary effect on the behaviour of India and Pakistan during the 2001–02 military stand-off and the subsequent crisis in 2008. The 1999 crisis also led to the articulation by India of its Draft Nuclear Doctrine and by Pakistan of its Nuclear Command and Control structure.

*Source: Naeem Salik, *Learning to Live With the Bomb*, pp. 11–12.

Experiential Learning[1] or Learning by Doing	This is the most common form of learning that can also be found in many other spheres of human activity and comes about with the experience of dealing with nuclear weapons over a period of time—a kind of on the job learning. The degree of learning will vary amongst individuals as well as institutions depending on the availability or lack thereof of an environment conducive to learning. Also, not everyone is likely to learn the same lessons from a similar experience. This type of learning is progressive and lessons learnt at an earlier stage of learning continue to be modified and improved upon with further experience, availability of new information, changes in the security environment, or technological developments.
Learning by Emulation or Imitative Learning	This type of learning entails learning from the experiences of other nuclear states, drawing useful lessons and adapting these in accordance with the peculiarities of one's own environment. This can be done with the help of publicly available information. For instance, nuclear command and control systems in the US, UK, and France were studied and useful ideas adapted to suit Pakistan's environment while designing the nuclear command and control. However, this kind of learning is limited to selected areas only because of sensitivity of the information and legal limitations imposed by domestic laws as well international treaty obligations. In most cases, the states with greater experience of handling nuclear weapons only share 'good practices' in areas such as safety and security; sensitive materials protection, control, and accounting (MPC&A); and export control and regulatory practices. US–Pakistan cooperation in security training, MPC&A, and export controls is a case in point.

Learning by Trial and Error	This is an innovative type of learning and can be very valuable provided an individual or an organisation is willing to accept mistakes and learn the right kind of lessons from these after due analyses, which requires a lot of courage to do. In the nuclear realm, due to the hazardous nature of technology, those dealing with this technology are generally not prepared to take any chances and therefore in the nuclear realm, this is not a popular kind of learning, although in many human endeavours this type of learning could be very productive.
Factual Learning[2]	In terms of nuclear learning, this type of learning refers to knowledge about the technical characteristics of nuclear weapons such as their yield and the effects a nuclear detonation produces in the form of heat, blast, and radiation. It is important for both the publics and the policy makers to be aware of the power of the nuclear weapons and their peculiar nature as distinct from conventional weapons. Such knowledge would ensure that the public would not exert undesirable pressure on the decision makers in times of crises and the leaders would know the consequences of any decisions they take with regard to the use of nuclear weapons.

Inferential Learning[3]	This type of learning involves the broader policy implications of the possession of nuclear weapons and it is important for policy makers to be aware of these implications for their security policy to enable them to make adjustments wherever necessary to adjust either the ends of the policy or the means employed to achieve those ends, or both. However, sometimes otherwise knowledgeable and experienced people fail to fully comprehend the implications of their decisions and actions. Henry Kissinger ordered DEFCON-III during the crisis arising out of the 1973 Arab–Israeli War without fully understanding its practical implications and technical details and its escalatory potential.[4]
Unlearning	In some cases useful and correctly learnt lessons are unlearnt. A case in point is Pakistan learning from the Cold War experience the destabilising potential of the battlefield nuclear weapons and eschewing their development but then deciding in 2011 to move down this path.

Notes and References

1. This term has been borrowed from Russell Leng. See, Russell J. Leng, *Bargaining and Learning in Recurring Crises: The Soviet-American, Egyptian-Israeli, and Indo-Pakistani Rivalries* (Ann Arbor: The University of Michigan Press, 2000).

2. This term has been borrowed from Jeffrey Knopf. See, Jeffrey Knopf, 'The Concept of Nuclear Learning', *Nonproliferation Review* 19, no. 1 (March 2012), pp. 79–93.

3. Ibid.

4. Richard Ned Lebow and Janice Gross Stein, *We All Lost the Cold War* (New Jersey: Princeton University Press, 1995), pp. 251–7.

1

A Brief Historical Overview of India's Nuclear Programme

Naeem Salik

Preamble

India has been one of the earliest entrants into the nuclear game thanks mainly to the foresight of Dr Homi Bhabha who had the benefit of his tutelage under Lord Rutherford, one of the foremost nuclear experts of the time. Bhabha had initiated the spadework for a prospective Indian nuclear programme as early as 1944 while India was still awaiting independence from the British Empire. This preparation enabled him to set up a nuclear energy commission in less than a year after Independence. Consequently, in the 1950s, India became a leading nation in the nuclear field amongst the Afro-Asian nations and by the late 1960s, it had acquired all the components of the nuclear fuel cycle. In contrast to the other nuclear powers that had initiated their respective nuclear programmes with the explicit purpose of achieving nuclear weapons capability, the Indian programme was initiated ostensibly to exploit the peaceful applications of nuclear technology. However, the programme was so designed that the option for eventual nuclear weapons capability was built into it. The first nuclear test in May 1974 clearly alluded to the ultimate goal of the programme, though the test christened the 'Smiling Buddha' was projected as a peaceful nuclear explosion. However, after at least two abortive attempts at retesting, India ultimately carried out a series of five tests—three on 11 May 1998 and two on 13 May 1998—and finally brought its nuclear bomb out in the open by declaring itself a nuclear weapons state.[1]

The nuclear explosion in 1974 was a landmark event because, from a technical viewpoint, there is hardly any difference between the configuration of a so-called peaceful nuclear explosive device and the one designed to lead to the production of a nuclear weapon.[2] What is important is the fact that India crossed the psychological threshold by demonstrating its readiness to face the likely international opprobrium and sanctions when it detonated a so-called 'peaceful nuclear device' on 18 May 1974. The international reactions were mitigated largely by India's insistence on the peacefulness of the explosion as well as the compulsions of international power play between the major powers. Nuclear detonations aimed at achieving fully operationalised military nuclear capability are logically followed by the development of appropriate nuclear delivery systems, and strategic command, control and communications systems based on a suitable nuclear use doctrine. In the aftermath of the 1974 test, India quietly started working on these essential requirements; however, the surreptitious manner in which these aspects were pursued helped keep international attention away. The multiple tests in 1998, however, removed any semblance of ambiguity. Coupled with this, the deliberate effort on part of the Indian leadership to provoke Pakistan finally brought the South Asian nuclear weapons programmes out of the closet.

The Indian Nuclear Programme from its Origin to 1974

The seeds of the Indian nuclear programme were sown on 2 March 1944, more than a year before the 'Trinity' test, a year and a half before the nuclear bombing of Hiroshima and Nagasaki, and over three years before the country's independence.[3] This unique but relatively little known aspect of the Indian nuclear programme can be attributed to the personal efforts of an enthusiastic young physicist, Dr Homi J. Bhabha, who had just returned after completing his studies at Cambridge University. Dr Bhabha managed to establish the Institute for Fundamental Research, with the assistance of the Sir Dorabji Tata Trust, for the expressed purpose of preparing a crop of Indian nuclear scientists and engineers. Bhabha was certainly much ahead of his time in visualising the peaceful uses of nuclear energy, at a time when the Manhattan Project in the US and other similar secretive efforts in some other countries, notably the UK and Germany, were focused only on the military potential of the atom. This is evident from the letter Homi Bhabha wrote to the trust declaring that, 'When nuclear energy has been successfully applied for power production in say a couple of decades from now, India will not have to look abroad for its experts but will find them ready at hand.'[4]

The Indian Atomic Energy Commission (IAEC) was set up within a few months of Independence with Bhabha at the helm. In the early 1950s, Homi Bhabha articulated the objectives of the Indian nuclear programme. He also laid down the timelines for the achievement of various goals during the first 25 years as under:[5]

- Achievement of technological self-sufficiency in the long run in various facets of nuclear research and development;
- in the short term, technology would be acquired from all available sources;
- development of designing and manufacturing capabilities along with the import of technology;
- breeder reactor technology would be developed in due course;
- large number of scientists and technicians would be trained both at home and abroad.

The high priority assigned to nuclear research and development is evident from the fact that the IAEC was under the direct supervision of Prime Minister Jawaharlal Nehru right from the outset. Its successor, the Department of Atomic Energy (DAE), established in

1954, also remained under the personal charge of the Prime Minister, which not only enabled Bhabha to evade the bureaucratic red tape but also allowed direct access to the chief executive of the country. Within two years of the establishment of the DAE, Bhabha led a team of Indian scientists who successfully carried out the first sustained chain reaction in Asia, at the indigenously constructed 1 MW research reactor APSARA. The enriched uranium fuel for the reactor was procured from Britain.

In 1958, Bhabha claimed that India could produce a nuclear explosive device within eighteen months of the political decision to do so.[6] This was clearly an exaggerated claim given the fact that at the time, India had no source of fissile material—the first plutonium production reactor CIRUS became operational in 1960 and India had no reprocessing facility at the time. More significantly, the claim was made six years before the first Chinese nuclear test in October 1964 and even before the Sino-Indian border conflict of 1962, which renders the narrative of the Indian nuclear programme as a response to the Chinese nuclear test incredible.

The Canadian supplied CIRUS reactor—a 40 MW capacity reactor that came online in 1960[7]—used natural uranium as fuel and heavy water as moderator and was thus eminently suited for production of weapons-grade plutonium especially if the fuel was removed at low burn up. This reactor could produce up to 15 kg of plutonium working on 100 per cent capacity. However, running on the usual operating capacity of 60–80 per cent, it could produce 9–12 kg of plutonium annually. Declassified US government documents confirm that Indians had indeed been removing fuel from CIRUS at low burn up.[8] It remained an important source of India's fissile material inventory until its retirement in 2010.[9] India's third research reactor, Zerlina, with a capacity of 100W became operational in 1961.[10] This reactor, which was indigenously designed, engineered, and built, was decommissioned and dismantled in 1983. In 1962, India's first heavy-water plant was commissioned at Mangal. In 1963, an agreement was signed with the US for assistance in the construction of a nuclear power station comprising two nuclear power plants at Tarapur.[11]

There is a general perception that from 1958 till the early 1970s, when the decision to conduct the PNE was ostensibly taken, the Indian leadership did not give in to the demand of Bhabha and the scientific community for the production of nuclear explosive devices. This belief in the narrative of the peaceful intent of the Indian nuclear programme can be found in most analyses by US government agencies—including the intelligence agencies—from the late 1950s to the early 1970s. A scientific intelligence report on the Indian Nuclear Energy Programme, prepared by the Central Intelligence Agency (CIA) in 1958, declared that there was no evidence to suggest India's interest in a military nuclear programme and its focus was totally on peaceful uses. The conclusion drawn by authors of the report was clearly influenced by their belief in India's traditions of 'passivity and mediation'.[12]

The Sino-Indian border war of 1962 made the US empathetic towards India and this feeling further deepened after the Chinese nuclear test in 1964. Consequently, the US tried to reassure India of US support in the event of a Chinese nuclear threat to India. These assurances were aimed at dis-incentivising India from going nuclear. However, if India decided to go nuclear, the US would show an understanding of its compulsions and take a benign view of India's actions. Efforts were made, including sharing of classified US data on the costs of maintaining a nuclear weapons capability, to dissuade India from going nuclear given its huge costs that would encumber India's weak economy.[13]

The American Consulate at Mumbai reported the inauguration of India's first reprocessing plant in April 1964 in a positive light by describing plutonium as a 'fuel for future reactors' rather than a potential bomb material. This development, which had made India the fifth country in the world besides the USA, USSR, UK, and France and the only one in Asia to have that technology,[14] did not cause any anxieties or result in international censure as has been the norm since the 1970s. When the construction of this plant began in 1963, the Indian scientists were fully aware of the dual purposes for which the separated plutonium could be used.[15] It was immediately employed for the separation of plutonium from the spent fuel generated by CIRUS.[16]

During this period, American diplomats and officials gave friendly advice to their Indian counterparts rather than any threats of sanctions or deprivation. In a brief for Governor Harriman on the eve of his visit to India in February 1965, the State Department declared with conviction that India had decided against the development of nuclear weapons and termed it a 'commendable effort worthy of emulation around the world'.[17]

The Americans were also careful in ensuring that neither their statements nor actions supportive of India should, in any way, affect India's non-aligned status or its political standing amongst the Afro-Asian countries.[18] However, it may be acknowledged that in the pre-NPT era, international norms, treaties, or agreements against nuclear testing or development of nuclear weapons by any state were non-existent and nor were there any technology denial groupings such as the NSG, MTCR, and the Australia Group.

The CIA in an October 1964 assessment stated that India has all the components to produce a nuclear weapon in a short time. However, it clarified in India's defence that it had no plans to initiate a bomb project yet because, in their opinion, the Indian government considered Chinese operational nuclear capability at least five years away. But if and when it materialised, India would be relying on President Johnson's assurances to come to the aid of all nations threatened by China.[19]

A scientific intelligence report prepared by the CIA in November 1964, however, identified the actual potential of India's nuclear complex stating that the Indian nuclear energy programme initiated in 1954 focused on peaceful purposes. It pointed out that India had three operational research reactors, one of which was an unsafeguarded reactor that could produce weapons-grade plutonium. India also had enough uranium fuel as well as a plutonium separation plant. A plutonium metal conversion plant was under construction and would be operational in 1966 or even earlier.[20] The report pointed out that India had imported unsafeguarded uranium concentrate from France, Belgium, and Spain. It also stated that fuel from CIRUS was being removed at low burn up, a clear indication that the reactor was being run in a manner to produce weapons-suitable material. The report mentioned that India's inaugural plutonium reprocessing plant had already processed 40 tons of spent fuel from CIRUS between March and August 1964. The report added that India had already accumulated 20 kg of plutonium nitrate. India had also requested Sweden to help design a pressurised heavy-water reactor with a capacity of 200–300 MWe. This reactor would be similar in design to the Canadian-supplied ones being built in Rajasthan but unlike the Canadian and US supplied reactors at Tarapur, it would be unsafeguarded.[21] This evidence was enough for any expert to recognise the true purpose and scope of India's nuclear activities.

Barely a year later, a special national intelligence estimate (SNIE) of the US on India's nuclear weapons policy concluded that India had the capability to produce nuclear weapons. It already possessed enough plutonium for a nuclear device which could be tested within a year of the political decision to go ahead; and not only had the bomb lobby gained strength from the September 1965 Indo-Pakistan war, but Prime Minister Lal Bahadur Shastri—an opponent of the nuclear weapons programme—had also boosted his political stature, which had strengthened his ability to oppose the military nuclear programme. The estimate, however, reached a grim conclusion that India would not pursue the nuclear abstention policy for long and was likely to detonate a nuclear device within the next few years and proceed further with weapons development.[22] The report confirmed that the CIRUS reactor has been operated in a manner so as to yield weapons-usable plutonium, and added that although a plutonium metal production plant is likely to be completed in 1966, a pilot reprocessing plant with sufficient capacity to process the plutonium produced by CIRUS was already working.[23] The intelligence estimate also surmised that there was a possibility that R&D related to weapons technology may already have been initiated. The estimate suggested that by 1970, India would have the capability to produce around a dozen 20 kilo-ton bombs and if the plutonium being produced by unsafeguarded power reactors was also dedicated for this purpose, it would substantially boost India's weapons production capacity.[24] The NIE, however, assumed lack of interest in nuclear weapons on part of the Indian military, which in its opinion would be more concerned about enhanced resources to augment its conventional capabilities.[25]

Contrary to his anti-bomb image, Prime Minister Lal Bahadur Shastri in an address to the Rajya Sabha (Upper House of the Parliament) on 16 November 1965 raised the possibility of India's development of nuclear weapons in response to the Chinese nuclear capability, stating that:

> While India stood for non-proliferation of nuclear weapons, if China developed her nuclear power and perfected the delivery system, then we will certainly have to consider as to what we have to do.[26]

However, he contradicted his earlier statement in his 3 December 1965 proclamation in which he stated that India had abandoned the idea of producing nuclear weapons because of its cost effects.[27] It is hard to judge whether this signified a return to his longstanding conviction or whether it was merely designed to assuage the concerns of the US officials, especially the reference to the economic costs involved.[28] However, the Indian nuclear enclave continued its work for the development of nuclear weapons capability in anticipation of a favourable political decision.

The concerns in the US State Department were growing as is obvious from a March 1966 telegram addressed to the US embassy in

New Delhi, pointing at the removal of fuel from CIRUS at less than 50 per cent burn up compared to its designed capacity. Adding that while this step alone may not be sufficient to conclusively determine that a decision has been taken to weaponise, it does imply a deliberate effort to create an option to actualise such a decision as and when it is taken.[29] The cable thus began by stating that, 'Although there is no evidence that India has decided to develop nuclear weapons, a nuclear device could probably be ready for testing within a year following such a decision.' It conceded that critical components like high quality detonators and electronic neutron generators likely to be used in a potential Indian nuclear device are readily available in the European markets. The State Department, therefore, asked its embassy to observe and report back any information on specific technical aspects such as activities in remote areas indicating construction of a nuclear test site, secret development of nuclear facilities or heightened security at existing sites, continued running of CIRUS in a way suited for production of weapon-usable plutonium, procurement of neutron generators and advanced high explosive detonators, or testing of shaped charges etc.

This was a far cry from the strong belief in India's peaceful intent, and serious apprehensions emanating from intelligence reports about India's nuclear weapons related activities were apparent. Such concerns were evident in the communication to the US embassy as well as the memorandum prepared by Acting Secretary of State George Ball for a National Security Council meeting in 1966 which stated that, 'India is almost certain to develop nuclear weapons,' adding that, 'at best we can cause a short term delay in India's plans.'[30] The Indian scientists on their part were also trying to convince the Americans that they were amenable to the American rejoinders about the economic costs of a military nuclear programme. Dr Sethna stated that, 'India cannot just detonate one or two devices and stop. [A] Small nuclear bomb program is worse than no program at all because it would invite [a] pre-emptive Chinese attack.'[31] He estimated that India would need

150 bombs for a credible deterrent.[32] Dr Vikram Sarabhai, the newly appointed secretary of the Indian Department of Atomic Energy and Chairman of the IAEC, seemed to concur with Sethna's views by acknowledging in a June 1966 statement that:

> A prototype bomb would not be useful as a weapon and India could not become an effective nuclear power without developing its industrial and economic potential ... mere possession of a bomb without a delivery system and a strong industrial base is a bluff.'[33]

Sarabhai's statement must have appealed to American sensibilities since they had been trying to persuade the Indians with similar arguments.[34] Acting Secretary of State George Ball explained in a memorandum for the US president that the third nuclear test by China had increased domestic pressures in India to embark on a military nuclear programme. He argued that, 'A pressure point is likely to be reached within a few years and unless there is some new development, India almost certainly will go nuclear.'[35] The memo also underscored the effect of such a step by India on Pakistan, which might seek Chinese or US help in its own development of nuclear weapons or by asking for a security umbrella to dissuade any Indian threat. Besides Pakistan, countries like Japan, Israel, and Germany may also follow suit. A nuclear India's dependence on American or Soviet assistance against a Chinese threat would also be reduced. The memo argued that threatening or imposing economic sanctions against India would reduce US leverage and push India closer to the Soviet Union. This will also amount to undermining a democratic state that is a counterweight to China.[36] Similar arguments are being offered these days to build India as a strategic counterweight to 'manage the rise of China'. India was even expected to give up its non-aligned status if it felt that its national security was at stake.[37]

The memo also suggested a joint American–Soviet guarantee to all non-nuclear states including India. However, since the Soviets were not amenable to this proposal, alternately, it was suggested that the

US should publicly pronounce its willingness to join other nuclear weapon states to give a nuclear guarantee to India or provide US assurance to India through a resolution passed by the UN. Chester Bowles, the US Ambassador to India, went even farther, proposing:

- US help to India in deploying an effective early warning system and other defensive systems to counter Chinese bombers.
- Extending the scope of combined Indo-US efforts for the surveillance of Chinese nuclear and missile developments.
- Surreptitious technical discussions about ballistic missile defence.
- Covert studies to develop joint air defences against possible Chinese nuclear attack that could involve the option of building an Indian manned bomber force capable of attacking Chinese missile launching sites [38]

During the course of these discussions, Americans were always sensitive to India's non-aligned status and did not contemplate a formal alliance with India which would not only have caused irreparable damage to Pakistan–US relations but could compel Pakistan to seek a military alliance with China. One of the extreme ideas considered entailed 'nuclear sharing' with India to boost India's ability to deter a possible Chinese nuclear attack with the help of American supplied nuclear warheads atop Indian delivery systems. Negotiating such an agreement would be complicated because of India's perceived reluctance to give up its non-aligned status. An arrangement of this nature with India would also negatively impact America's Asian allies, including Pakistan and Japan. Bowles, however, believed that it was a price worth paying. He also anticipated a possible Indian PNE to demonstrate its technological prowess though such an explosion would be perceived by Pakistan and other states as the beginning of an Indian military nuclear programme.[39]

While the State Department was exploring these ideas, the US Department of Defense thought otherwise. In January 1967, the Joint

Chiefs of Staff in a memorandum for the Secretary of Defense voiced strong differences of opinion with the State Department proposals and suggested instead that:

- No nuclear assurances be extended to India beyond those made by the President in 1964.
- No action be taken in regard to India which could alienate US allies, especially Pakistan.
- The United States avoid creating any impression that it is willing to broaden its commitments to India.
- The United States retain maximum flexibility for future US action in response to CHICOM (Chinese Communists) nuclear attack or blackmail.[40]

In the Joint Chiefs of Staff's opinion, despite serious US concerns with India's security, there was no justification for pledging a nuclear umbrella to India similar to the one given to the NATO states. They rebuffed the provision of ABM protection to India and rightly concluded, based on available intelligence, that despite all this assistance, India would conduct a nuclear detonation within the next few years.[41]

In April 1967, L. K. Jha, secretary to the Indian Prime Minister Indira Gandhi, came as her special envoy to meet President Johnson to seek NPT related security guarantees after having obtained an affirmative answer from the Soviets. Johnson gave an assurance to Jha that they would seriously consider the Soviet Draft and pursue it further.[42] In June 1967, the US Secretary of State Dean Rusk and Soviet Foreign Minister Gromyko discussed the issue of security guarantees. The Soviets asked for a matching US declaration to the Soviet pronouncement; Rusk expressed the inability of the US citing constitutional restrictions and insisted on a UN Resolution. In Gromyko's opinion, such a declaration could only be in the context of the NPT.[43]

Cognizant of the increasing Indian interest in PNEs, the Americans sent an aide-memoire to the Indian Atomic Energy Commission on 16 November 1970 that clearly articulated that the US did not distinguish between a PNE and a nuclear weapon and that a nuclear device produced even for a peaceful purpose could be used for destructive objectives as well. US' obligations under the NPT and the US Atomic Energy Act did not allow for any other interpretation.[44] The aide-memoire also reminded the Indians that the heavy water sold for CIRUS was meant for peaceful uses only, implying that any use of plutonium produced by CIRUS for non-peaceful purposes would be a violation of this agreement.[45] This particular document became a source of contention during the course of negotiations for the US–India civil nuclear agreement wherein the State Department officials were trying to exonerate the Indian violation of the agreement in 1974 by stating that the agreement was vague and could be interpreted in different ways. However, those opposed to the agreement referred to the aide-memoire as having dispelled any ambiguity whatsoever.[46]

India entered into collaborative agreements in the nuclear field with other countries as well. In March 1972, it signed a protocol to receive financing worth 35 million francs from France and a supply of nuclear pumps worth 7 million francs in addition to the Swedish-made materials for the Madras Atomic Power Project.[47]

India's First Nuclear Explosive Test

India shocked the world and especially its immediate neighbours by conducting a nuclear explosion on 18 May 1974, though it tried to mitigate its impact by terming it a PNE. The plutonium used for this explosion was extracted from the CIRUS reactor's spent fuel in contravention to the understanding with Canada and the US who had supplied the technology and materials for the plant. The underground explosion, carried out at a depth of 100 meters, created a crater of 150 meters in diameter. The designed yield of the device was claimed to be 10–15 kilotons; however, many experts around the world concluded that the actual yield was far less than the Indian claims. The IAEC's official statement characterised the detonation as a PNE, hinting at a plan to explore the peaceful uses of nuclear detonations, especially in areas such as mining and earth-moving projects.[48] Later events were to prove that such statements were only designed to assuage the international pressure and no such project was undertaken thereafter.

This theme of peaceful tests was repeated time and again by the IAEC which in fact expressed its revulsion to military uses of nuclear technology. The Indian Prime Minister, Indira Gandhi, on her part reiterated the peaceful nature of the explosion and tried to assuage the concerns of India's neighbours by asserting that India did not want to use nuclear technology for any purpose other than peaceful applications. The defence minister in an interview with journalists the day after the test also ruled out the possibility of India using nuclear technology for manufacturing nuclear weapons, while the foreign minister, shrugging off any intent to make nuclear weapons, dubbed the nuclear test an effort towards harnessing nuclear energy for peaceful and economic purposes.[49] These professed peaceful objectives, however, were not convincing because technology used for both PNEs and weapons is fundamentally the same, something the US aide-memoire had also made amply clear.[50] India had become, in reality, a de facto non-NPT nuclear weapon state.

The technical configuration of the Indian device was also indicative of the in-built weapons potential. Onkar Marwah has alluded to some special characteristics of the Indian nuclear explosive test as under:

- Initiation of the nuclear explosive programme with an underground test, which no other nuclear power had done at the early stages of development.

- Implosion mechanism was used which is a sophisticated technology with possible weapons applications.
- Claimed yield of the explosion at 10–15 kiloton suggested the size of a nominal yield fission bomb.
- Special safety features were incorporated to ensure a 'clean explosion'.[51]

The Indian government made every effort to convince the world that its explosive test was indeed a PNE designed to exploit nuclear explosive technology for purposes such as mining and digging of canals etc. However, not even the established nuclear powers had been able to employ nuclear explosions for such purposes and there is no evidence to suggest that these methods are viable. A well-known Indian strategist, K. Subrahmanyam, argued in support of the 'PNE theme', explaining that the Indian scientists as well as the government leaders know that it will be years before India could master the nuclear explosive technology for peaceful purposes as would be the case for the US and USSR. Echoing Homi J. Bhabha, he added that when this technology becomes normal to apply for peaceful uses, India should not be dependent on any outside assistance and has therefore taken a start now.[52]

India's effort to defend its nuclear explosive test was at cross-purposes with its traditional stance against nuclear explosive testing at various international forums. Prime Minister Nehru had himself asked the UN General Assembly to put a stop to all nuclear tests with a view to prevent the proliferation of nuclear weapons. India had also been an ardent supporter of a Comprehensive Test Ban Treaty (CTBT). As far back as August 1965, the Indian ambassador to the Eighteen Nation Disarmament Committee (ENDC) had stated that India considers all nuclear tests as evil and had asked for this evil to be addressed as soon as possible.[53]

International Community's Response to the Indian Test

International reactions to the Indian test were mixed. The US had made it known to India that it did not make any distinction between PNEs and weapons tests; however, its reaction to the Indian test was understandably mild. Henry Kissinger in a conversation with his Canadian counterpart had remarked that there is no use fighting a fait accompli.[54] The Soviets on their part did not want to push India into the US orbit and refrained from criticising the Indian action. As pointed out by the US mission at NATO, though the Soviets were worried about proliferation, they did not want to spoil their relations with India and as such, emphasised on the peaceful nature of the Indian test.[55] The American and Soviet reaction was understandable because of their hostility towards China. They did not want to portray the Indian test as a negative development and in fact thought that a nuclear-armed India would be better equipped to challenge China.

The Chinese avoided making any public comments so as not to dignify the Indian action by any official reaction. However, the Chinese were expected to carefully analyse its impact on not only the bilateral India–Pakistan ties but on the process of normalisation of their own relations with India as well.[56] There was widespread condemnation of the Indian nuclear test amongst the Japanese politicians as well as the Japanese media. However, the Indian action was condemned in the strongest terms by Canada and Pakistan. Some in the West thought that a poor country like India could not afford the costs of such a venture but in reality, India did not feel the economic burden since it had derived the wherewithal for this test from its long-standing peaceful nuclear program. The actual cost of the explosive device was therefore insignificant. Indian government estimates of the expenses on plutonium and preparation of the test site stood at merely USD 400,000.[57]

In the aftermath of the test, the US cut off the supply of enriched uranium fuel for the nuclear power plants built at Tarapur with US assistance. Interestingly, the non-proliferation champion, Jimmy Carter, renewed the supply. It was discontinued again by the Reagan administration. However, the US encouraged France to start the fuel supply to keep the plants running. The Soviets on their part provided the heavy water for the Canadian-supplied reactors when the Canadian government cut off the supplies in 1976.

A *Washington Post* article summed up the views of those who denounced the Indian test explosion as under:

> India's peaceful nuclear explosion experiment is, first of all, the test of a bomb. Not only is there no real distinction between a military and peaceful explosion but even the United States with all its time and technology has yet to find a single feasible peaceful use for nuclear explosives. For India to call its explosion 'peaceful' and to abjure all military intent is in a word rubbish. ... The fact is that India, which has long had the capability to do so, has now gone nuclear in the political military sense. ... the Indian explosion is the height of irresponsibility. Whatever the supposed gains in national pride and governmental prestige and regional political standing, the blast can only further aggravate Pakistan's fears of Indian domination and slow the normalisation process that had been unfolding recently in the South Asian subcontinent. In a wider orbit the Indian test will in effect license and strengthen various other countries—Japan comes quickly to mind—the internal forces partial to building national nuclear bombs.[58]

The Canadian response was the strongest since it felt aggrieved and betrayed, having been the major source of technology as well as funds for India's ostensibly peaceful nuclear research and development. It was also known that the plutonium used in the explosion was taken out of the CIRUS reactor supplied by Canada for peaceful purposes and the Canadians felt that the Indians had violated the spirit of the bilateral agreement. Canadian Foreign Minister Mitchell Sharp articulated the anguish of the Canadian government in a statement on 22 May 1974, emphasising that Canada does not make a distinction between peaceful and military explosions and is concerned about the negative impact of the Indian action on international non-proliferation efforts. He also expressed his country's dismay over the fact that it had given substantial assistance to India in the nuclear energy domain and although India is free to take its decisions, Canada cannot become a facilitator for a nuclear program whose objectives are at cross-purposes to Canadian policy.[59]

Prime Minister Zulfikar Ali Bhutto, in a hard-hitting statement, emphasised that Pakistan would never be overawed by the threat posed by the emerging Indian nuclear capability. He announced that the Foreign Secretary was travelling to Beijing, Paris, and London to seek guarantees against any use of nuclear threats by India while he would personally take up the subject with the Soviet leadership as well as the Canadian officials. He added that the Minister of State for Foreign Affairs and Defence, Aziz Ahmed, had been asked to broach the matter not only at the CENTO meeting due to be held in Washington but also with the US officials. Bhutto also ruled out the acceptance of India's proposed 'No War Pact', terming it a submission to blackmail.[60]

India on its part was dismissive of the criticism as epitomised by the response of K. Subrahmanyam, Director of the IDSA, and a strong advocate for an Indian bomb. In his address to the Indian International Club on 1 August 1974, he clearly alluded to the real objectives of the PNE, thereby unveiling the falsehood of the peaceful purposes of the test claimed by the Indian officialdom. The following excerpts from his talk make the point amply clear, pointing out that the NPT makes a distinction between peaceful and military nuclear explosives. On the question of credibility (of the Indian declaration of peaceful purposes), he stated that all declarations in international politics are dependent on the situation and as there are no permanent friends or enemies, there are no permanent policies but only permanent objectives.[61] His remarks exposed the thin veneer over the official Indian pronouncements about the PNE.

Subrahmanyam's arguments leave no doubt that what kept India from pursuing a military nuclear program were not any pious sentiments but certain technological deficiencies. These included the lack of unsafeguarded plutonium production reactors, the inadequacy of technical wherewithal to establish a nuclear command and control, lack of requisite nuclear delivery systems, absence of an appropriate doctrine, and an underdeveloped electronics industry. He also argued that India did not need nuclear capability against Pakistan due to its conventional advantage vis-à-vis Pakistan, while a few weapons would not be effective against China or against a nuclear threat from a super power.[62] He conveniently papered over the fact that CIRUS had already produced plutonium worth 10–12 weapons by 1974 and would continue to be an important source of fissile material until its shutdown in 2010. Dr Vikram Sarabhai on his part, made an effort to rationalise the so-called PNE by using the argument of India's primitive economy, which had led to its colonisation, highlighting the need for progress in advanced industrial sectors such as electronics, aerospace, nuclear, computers, and automation.[63]

The prestige factor was clearly discernible in the reactions of the political classes and the media. The media advanced the arguments of the government with regard to the peaceful nature of the test. The populace at large and the media seemed to be convinced that the test would raise India's status in the international community, establish India's credentials as a country capable of effectively harnessing available resources, and strengthen the case for removing the inequalities inherent in the NPT.[64] However, some members of the ruling party and senior military officers were convinced that the peaceful purposes argument was spurious and India would eventually develop nuclear weapons capability. Nevertheless, Indian diplomats were concerned that in case India failed to demonstrate peaceful applications of the PNE, its credibility would suffer. The hardliner Bharatiya Jana Sangh party (the predecessor of the present-day Bharatiya Janata Party or BJP), on its part, ostracised Prime

Minister Indira Gandhi for ruling out the weapons option and urged the government to pursue it.[65] An American estimate based on the domestic reactions suggested that any imposition of sanctions would further solidify public support for further nuclear testing.[66] The American assessment typifies the self-created fears of Indian annoyance that were used to justify tolerance of Indian transgressions. Such an approach emboldened India to pursue its nuclear objectives without fearing adverse consequences.

It is difficult to identify the factors that led to the Indian decision to go ahead with the nuclear test and its timing. The general perception is that the decision was taken some time in 1972. The Indian official narratives tend to overplay the Chinese and American role during the Bangladesh crisis in 1971. The facts however, do not support this argument. The Chinese never intended to intervene militarily on Pakistan's side and their options were further constrained by the Treaty of Friendship and Cooperation—a virtual defence cooperation agreement—between India and the USSR, signed in August 1971. As for the US, the episode involving the movement of the Enterprise Task Force towards the Bay of Bengal is also exaggerated since it neither intervened on Pakistan's side nor played any role in evacuating the Pakistani forces from erstwhile East Pakistan. In fact, it had a deleterious effect on the morale of Pakistani soldiers who were vainly hoping for its intervention. At the same time, the Soviet Union not only provided weaponry to India but obstructed all efforts at the UNSC to bring about an early ceasefire, thereby gaining time for the Indian military to bring its operations to a conclusion. Despite the much-touted American tilt in favour of Pakistan, the US administration did not take any practical steps to assist it, such as by lifting the embargo on military supplies. It even turned down a Jordanian request to be allowed to transfer four F-104 fighters as a goodwill gesture that in any case would not have had any impact on the military situation.[67]

Indian analysts and government officials frequently bring up US intervention in the 1971 war by sending its naval task force in support of Pakistan but would never make a mention of the fact that a similar naval task force was sent to the Bay of Bengal on Prime Minister Nehru's request during the Sino-Indian conflict of 1962.[68] The 1971 war should actually have boosted the confidence of the Indian leadership—if the US did not do anything to prevent the dismemberment of its long-time ally Pakistan, it is unlikely to interfere in any future India–Pakistan conflict. The limits of Chinese ability to interfere in an India–Pakistan conflict were also laid bare.

Rodney W. Jones, a renowned expert of South Asian nuclear affairs, while analysing India's decision to explode a nuclear device commented that:

> Authorisation to prepare a nuclear explosive device may have been proposed and considered earlier but was probably not granted until late 1969 or early 1970. Reports at that time of a successful Chinese long-range missile test provoked fresh pressure from Parliament. The split of December 1969 in the ruling Congress Party probably gave Indira Gandhi an incentive to seize the initiative and remove the nuclear issue from inner circle contention. About the same time in May 1970 Vikram Sarabhai announced a ten-year programme (the so called 'Sarabhai profile') to accelerate nuclear energy and space technology development.[69]

If one were to believe Jones' assessment, the linkage between the decision to test and the 1971 war peddled by the Indians has no factual basis and the decision was probably rooted in either the domestic political compulsions or advances in Chinese missile capabilities. It may be recalled that in the immediate aftermath of the first Chinese nuclear test in October 1964, Bhabha had reasserted his longstanding claim that India could fabricate its own bomb within eighteen months. Going by Bhabha's claim, India could have produced a bomb by mid-1966. However, that plan was set back by the deaths of Prime Minister Shastri and Dr Bhabha within a few months of each other in early 1966.

In an alternative argument based on the Indian viewpoint as he himself points out, Jones averred that:

> The one development between late 1971 and the following year that might have served as a specific stimulus to initiate (or rather confirm) a project for a nuclear explosive test would have been evidence which Indian intelligence probably picked up that Z. A. Bhutto had deliberately launched an active programme in Pakistan for the development of a nuclear weapons capability shortly after he assumed control of the government. *This would coincide with (and perhaps explain) some Indian opinions conveyed to the author that the official decision to prepare for the nuclear demonstration occurred in the early autumn of 1972* [emphasis added].[70]

It appears that the motivation for the decision to test came from Indira Gandhi's mounting domestic political troubles as is evident from a CIA analysis of the test which declared that:

> The decision to go ahead at this time was probably made in order to boost India's sagging international prestige and to divert public attention from the government's mounting domestic problems ....[71]

A report prepared for the CIA director assessing the reasons for failure of the intelligence community to predict the Indian nuclear test concluded that:

- The intelligence community had failed to warn US decision makers of the planning and preparations for the test, denying them the opportunity to pre-empt this important proliferation activity.
- Though the intelligence agencies were aware of India's capability to fabricate and test a device and had made assessments as far back as 1965 that India was destined to explode a nuclear device

in the next few years, yet they failed to foresee the actual event perhaps due to the fact that they had not accorded it a high enough priority and lack of coordination and communication between various elements of the community itself.

• It was essential for the intelligence agencies to assign a much higher priority to this task than they had done so far.[72] This soul searching, leading to according of higher priority, helped the intelligence community in the timely detection of India's preparations for another test in 1995, thereby allowing the US government to use its diplomatic clout to force [the] Narasimha Rao government to call off the test.[73]

Post-1974 Developments

After the first Pokhran test, the Indian government insisted that it did not have any desire to develop nuclear weapons. However, it never ruled out the possibility of further PNEs for R&D purposes. It would be over two decades before India would conduct more nuclear tests. In the intervening period, India kept its nuclear plans low key and shrouded in ambiguity. In reality, India continued to pursue its nuclear ambitions, paying greater attention to associated technological areas such as nuclear delivery systems, space capabilities including surveillance, and communications and electronics, etc. In 1975, it also started construction of its second and largest plutonium production reactor named Dhruva or R-5 with a capacity of 100 MW. This reactor went critical in 1985.[74] It is evident that India wanted to have all the components of an operational nuclear capability available in the event it decided to make its nuclear weapons capability public. The ominous direction of the program became even clearer with the commencement in 1983 of the Integrated Guided Missile Development Programme (IGMDP) aimed at producing five different missiles including Prithvi, a short range missile, and Agni, an intermediate range ballistic missile—both capable of carrying nuclear payloads.[75]

India's space programme was substantially enhanced in the early 1970s and was closely integrated with the nuclear program as part of the ten-year development plan announced by Vikram Sarabhai. In July 1974, a few months after the nuclear test, the head of the Indian Space Commission pronounced that India now had the capability to develop medium range missiles using indigenous solid fuel and guidance systems. In the next year, India launched a research satellite from a Soviet launch site.[76] Parallel to the nuclear and space programmes, the entities dealing with electronics were also given tasks and schedules matching those of the nuclear and space programs.

However, in spite of some success in producing short as well as intermediate range missiles, India failed to pronounce its status of a nuclear weapons state. Some experts opine that the Indian nuclear venture had been beset with several technological difficulties ever since its first nuclear test. The problems were caused by cessation of Canadian assistance in response to the Indian test explosion as well as restrictions on export of nuclear-related technology applied by other supplier countries. On the other hand, India's indigenous development of critical materials and technologies was not keeping pace with the targets. Consequently, India was compelled to sign an agreement with the Soviets in 1976 for the supply of heavy water under strict IAEA safeguards that in turn brought any facility using that material under IAEA safeguards and inspections regime. India, which had always been reluctant to open its nuclear facilities to IAEA inspectors, had to swallow this bitter pill to keep its plants running. It also meant that the possibility of diversion of spent fuel for extraction of fissile materials from these plants was also foreclosed.[77]

An Indian analyst, Bhabani Sen Gupta, acknowledged that export restrictions imposed by the London Suppliers Club seriously impeded India's peaceful nuclear programme. The components required for a military program were banned in any case. The export controls delayed the completion of the Dhruva reactor, which was vital for the pursuit of a military programme. Several mishaps had also delayed the completion of India's heavy-water plants without which Dhruva could not be operated as an unsafeguarded reactor. This had convinced the Indians that foreign powers were deliberately trying to undermine and retard India's nuclear programme.[78] He thought that India could acquire nuclear weapons by the late 1980s at the earliest but this could be delayed until the mid-1990s.[79] Dhruva was finally inaugurated in 1983, five years behind schedule.[80]

The Reality of the Chinese Threat

In the aftermath of its border war with China in 1962 and especially after the Chinese nuclear test in 1964, India has been projecting China as the primary threat to its security and Pakistan as a secondary threat. However, despite the unresolved disputes, the relations between India and China, especially the economic and trade relations, have significantly improved. India, however, expanded its nuclear R&D programme on the premise of a potential nuclear danger from two nuclear-armed neighbours. India's endeavours in this regard were not only bound to offend the US and other industrialised states but would also reverse the trend of improvement of its relations with China. While India was loathe to be equated with a smaller Pakistan, it could not ignore any Pakistani advances in the nuclear domain. It, nevertheless, continued to propagate the virtues of PNEs. This provided a convenient cover for India's nuclear weapons-related activities without any practical demonstration of PNEs.[81]

In the beginning of the 1980s, India started diversifying its nuclear fuel cycle by commencing research on uranium enrichment technology, which included experimentation by Bhabha Atomic Research Center scientists on Laser Isotope Separation (LIS).[82] Raja Ramanna, the Chairman of the DAE, claimed in November 1986 that scientists at BARC had successfully accomplished uranium enrichment using gas centrifuge technology based on the 'Zippe' design.[83]

In the same month, a report by Delhi-based US Defence Advisor to the Defense Intelligence Agency (DIA) DIA) stated that Indian nuclear programme was plagued with technical glitches that included shortage of heavy water, difficulties in operating the Rajasthan Power Station, and frequent breakdowns of the Dhruva Reactor. At the same time, the domestic debate for or against the acquisition of nuclear weapons was also picking up pace. Apparently reacting to rumours about Pakistan's nuclear developments, Prime Minister Rajiv Gandhi told senior military officers that India had no intention to produce nuclear weapons but if it were faced with nuclear weapons across its borders, it would have to consider all options.[84]

The Soviet offer to provide light water reactors to India became a contentious issue because this was considered a deviation from the long-standing policy of indigenising heavy-water reactors and secondly, there was no domestic capability to produce enriched uranium fuel which raised the question of reliability of foreign fuel supplies. Dr Ramanna, though, declared that, 'India is in a position to produce as much enriched uranium as might be required to support these two reactors.'[85] However, this was certainly a tall claim since India had been unable to fuel the Tarapur plant and the availability of imported fuel was becoming increasingly difficult. This explained India's pronouncement that it was planning to modify the reactor design to enable it to run on Mixed Oxide Fuel (MOX) fuel. However, due to legal complications, India gave up this option. Two decades later, India continued to import fuel for the Tarapur plant from Russia to keep it operational.[86]

In December 1995, US satellites detected preparations for a nuclear test at Pokhran. The evidence was presented to the Narasimha Rao government and he was persuaded to call off the test.[87] This show-and-tell exercise taught the Indians to evade detection of their nuclear test preparations in May 1998. This was the major cause of what was termed by Richard Shelby, Chairman of the Senate Intelligence

Committee, as a 'colossal failure' on part of the CIA in timely detection of the event.[88]

In February 1998, the BJP unveiled its election manifesto in which it unambiguously stated its intention to test and induct nuclear weapons should it gain power. The party made a commitment that as part of its 'external security' policy it will:

- Establish a National Security Council to constantly analyse security, political, and economic threats and render continuous advice to the Government. This council will undertake India's first ever strategic defence review to study and analyse the security environment and make appropriate recommendations to cover all aspects of defence requirements and organisation.
- Re-evaluate the country's nuclear policy and exercise the option to induct nuclear weapons.
- Expedite the development of the Agni series of ballistic missiles with a view to increasing their range and accuracy.
- Increase the radius of power projection by inducting appropriate force multipliers such as battlefield surveillance system and air to air refuelling.
- Enhance the traditional and technical capabilities of our external intelligence agencies and also increase the interaction and coordination with user departments.
- Place paramilitary forces in sensitive border areas under the full control of the army.[89]

A Pakistani foreign ministry spokesperson, while voicing his disquiet over the BJP manifesto, declared that Pakistan would be compelled to revisit its policy to maintain its security and sovereignty. On 7 February 1998 Richard Celeste, the US Ambassador to New Delhi, stated that his country would have serious concerns if India declared itself a nuclear weapon state.[90] However, many analysts felt that this was nothing more than election rhetoric and if the BJP came to power, it would back away from this declaration. The BJP itself encouraged this line of thinking by toning down its stance on various political issues and its own nationalistic outlook and its mild reaction to Pakistan's testing of the Ghauri missile in April 1998. It was thought that the BJP would carry out a policy review through the envisaged National Security Council before undertaking any precipitate action. The hope that the BJP would wait until it received the NSC recommendations before taking any major decisions seems to have led the US administration as well as the intelligence community into a state of complacency, which in turn led to their failure to anticipate the timing of the Indian nuclear tests in May 1998.

The Shakti Series of Tests

A little over a month after taking the reins, the BJP made good its promise by conducting the Shakti series of tests. An Indian journalist reported that, 'Buddha today smiled for the second time at Pokhran, and grinned three times on his birth anniversary which coincided with India's triple nuclear test ... The triple tests included a simple fission device, a low-yield device, and remarkably a thermo-nuclear device.'[91]

Though Paul Laventhal of the Washington based Nuclear Control Institute believed that, 'This whole exercise is quite likely an exercise to showcase India's thermonuclear capability. It is the one test that India needed to do to show they have a Hydrogen Bomb',[92] the seismic analysis of the tests suggested a much lower yield than claimed by India and even raised serious doubts about the veracity of a thermonuclear test. The experts argued that while the primary stage of an attempted thermonuclear bomb may have worked, its secondary was a fizzle. Later, Indian scientists such as Srinivasan and Santhanam reconfirmed these doubts.[93] India claimed to have conducted two more tests of low-yield devices on 13 May 1998, which no seismological station picked up. One cannot be sure whether these tests were actually conducted but failed, were not carried out at all, or the yield was too low to generate any significant seismic activity that could be detected or measured.

The tests generated a strong nationalistic euphoria in India and a vast majority of people celebrated the event across the country. The post-test surveys suggested approval by 91 per cent of the population. The tests were viewed as a manifestation of India's growing technological prowess.[94] J. Robert Oppenheimer's famous quotation of a line from the Bhagavad-Gita, after witnessing the first nuclear test in 1945, 'Now I have become death, destroyer of the worlds' was recalled.[95] A former Indian foreign secretary, J. N. Dixit, declared that, 'the tests sent a clear signal that we are a confirmed nuclear power',[96] adding that the tests were in accordance with India's national security interests and would not only strengthen her regional political status but would also create a strategic equilibrium in the region.[97] Another former diplomat, S. K. Singh, rightly predicted that the sanctions resulting from the tests would be short-lived.[98]

In order to convince world leaders of the rationale for Indian nuclear tests, Prime Minister Vajpayee wrote letters to President Clinton and other G-8 leaders[99] in which he declared China as the primary threat to Indian security while also blaming Pakistan on several counts, stating that:

> We have an overt nuclear weapon state on our borders, a state which committed armed aggression against India in 1962. Although our relations with that country have improved in the last decade or so, an atmosphere of distress persists mainly due to the unresolved border problem. To add to the distress, that country has materially helped another neighbour of ours to become a covert nuclear weapon state. At the hands of this bitter neighbour, we have suffered three aggressions in the last fifty years. And for the last ten years we have been the victim of unremitting terrorism and militancy sponsored by it in several parts of our country, especially Punjab and Jammu and Kashmir.[100]

The Indian prime minister was clearly trying to appeal to American concerns about a rising China and also positioning India as a natural strategic counter to China. Defence Minister George Fernandez went a step further, calling China India's 'enemy number one'. Home Minister L. K. Advani and some other senior ruling party leaders started threatening Pakistan, reminding it of the changed strategic environment. These aggressive statements were meant to serve twin purposes of trying to browbeat Pakistan as well as daring it to respond by conducting its own nuclear tests that would mitigate international pressure over India and divert attention to Pakistan. These statements were termed 'foolish' and 'dangerous' by the State Department spokesperson who advised caution.[101] However, recognising the consequences of an open confrontation with China, the Indian leadership moved quickly into damage control mode.[102] A CIA intelligence report analysing the situation created by India's nuclear tests pointed out that:

- Although the BJP failed to win a majority in the general election, the party is riding a tidal wave of popularity from the tests and is now signalling that resolving Kashmir on India's terms will be next on its agenda.
- New Delhi is claiming that its nuclear tests were for national security and to counter China. Nonetheless, last week Home Minister Advani declared publicly that Pakistan must 'roll back its anti-India policy immediately' or 'it will prove costly' for Islamabad.
- Pakistan's decision to conduct nuclear tests is being portrayed by the BJP government as confirmation that its 'get tough' policy towards its neighbour was justified.[103]

Advani hinted at India's readiness to blackmail Pakistan on the Kashmir issue and warned Pakistan of serious consequences of any interference in Kashmir, reminding it of the change in strategic balance between the two countries after the Indian nuclear tests.[104] Advani talked of a more aggressive approach by the Indian military, including the possibility of hot pursuit operations.[105] The linkage

India was trying to create between its nuclearisation and the fate of the Kashmir dispute was demonstrated by the presence of the state Chief Minister Farooq Abdullah during Vajpayee's visit to the test site.[106] This belligerent posturing by Advani and other BJP leaders, however, disappeared soon after Pakistan's nuclear tests.

Hindu Nationalism and the Bomb

The decision to conduct the nuclear tests was used to create a nationalistic fervour and termed as a way of asking for recognition of the rightful place of the 'Hindu civilisation' in the international order.[107] It was also meant to be a vivid reminder to India's smaller neighbours to accept its hegemony in the region.[108] An Indian journalist pointed out, 'already, the first whisper of the Hindu Bomb is being heard. Many writers and analysts have attributed the tests as much to the BJP's domestic political compulsions and instinct for survival than India's security needs or strategic foresight. The result: A spate of editorials, analyses, and cartoons with a "Hindu touch".'[109]

The international response was weak and disjointed as was the case after the 1974 test. The US imposed sanctions mandated under the Nuclear Proliferation Prevention Act of 1994. The EU, Russia, and France, however, rebuffed any suggestions to apply sanctions. On the eve of a G-8 summit at London, its members asked Pakistan to accept India's pre-eminence. Apparently convinced by Vajpayee's reasoning, President Clinton stated his understanding of India's compulsions for carrying out the tests.[110] Not to be left behind, the speaker of the Russian parliament not only supported India's decision to test but also praised it as a correct decision in defiance of US pressure and acting on its national pride.[111] In a repeat of 1974, the world powers seemed to be resigned to the possibility of a nuclear India and appeared ready to accept its nuclearisation as a fait accompli.

Perceived Indian Designs for Going Nuclear

India's nuclear program was initiated for peaceful uses of nuclear technology. However, it was so designed that there was an inherent potential for the development of a military nuclear capability as well. Dr Homi Bhabha, the father of the Indian nuclear programme, himself led the bomb lobby until his death in a plane crash in 1966. The baton was then taken up by eminent strategic thinkers such as K. Subrahmanyam and some others in the Indian strategic community. Subrahmanyam had termed nuclear weapons as the currency of international power and had also attributed China's diplomatic recognition by the US as a consequence of Chinese nuclear capability.[112]

Beginning in the late 1970s, the Indian leaders had started using the pretext of a potential Pakistani nuclear option as a justification for the pursuit of a nuclear weapons option by India. Prime Minister Indira Gandhi declared on the floor of the Indian Parliament on 9 April 1981 that Pakistan's development of nuclear weapons capability would have serious repercussions for South Asia and declared that India would respond to any such eventuality in a befitting manner.[113]

Bhabani Sen Gupta went so far as to suggest that the focus of the domestic Indian nuclear debate in the early 1980s was solely on the likely development of a Pakistani military nuclear capability.[114] Contrary to the popular perception that India's nuclear weapons effort was in response to the Chinese nuclear test in 1964, Gupta argued that the Chinese test did not evoke the perception of a nuclear threat, nor did it influence India's decision to go nuclear. But that would not be the case in the event of a Pakistani nuclear weapons test or even a PNE, since it would be viewed as a direct challenge to India's regional dominance.[115] Indian Defence Minister C. Subramaniam, speaking at the National Defence College in October 1979, declared that he could not commit nuclear abstention by India in the future, listing the possible scenarios that would lead to India's nuclearisation, and pointed out that Pakistan's decision to go nuclear would top the list.[116]

On the other hand, Gupta explained that in addition to Pakistan's decision to go nuclear, other possible reasons for India to nuclearise could be India's sovereign decision to go nuclear, an act of defiance against an unfair non-proliferation regime, or even a delayed reaction to the Chinese nuclear weapons capability.[117]

However, the above justifications aside, the most critical element that emerges from the political statements as well as public perceptions has been the 'prestige' factor. There is sufficient evidence to suggest that the Indian nuclear establishment had made, from the very beginning, a deliberate effort to keep the nuclear weapons option open, and the political leaders went along. For instance, as far back as 1948, Nehru had told the parliament that, 'Of course if we are compelled as a nation to use it for other purposes, possibly no pious sentiments would stop the nation from using it that way.'[118] George Perkovich has cited a 1960 incident wherein a retired two star US Army General Kenneth D. Nichols, visiting India as a Westinghouse consultant discussed plans for the construction of the Tarapur reactor, and briefed the Indian Prime Minister. To Nichols' surprise, Nehru unexpectedly enquired from Bhabha, who was also present at the briefing, whether he could develop an atomic bomb and how much time he would require to do so. Bhabha responded in the affirmative, adding that it would take him a year. Nichols was perplexed to hear Nehru, reputed to be a peace icon, asking such questions. Nehru even asked Nichols' opinion about Bhabha's claim to which he replied that he had no grounds to dispute his claims. Nehru ended the conversation by asking Bhabha not to do it until he directed him.[119]

Indira Gandhi reportedly took the decision for the first nuclear test in 1972 when, after her triumph in the 1971 war, she was at the height of domestic popularity and was enjoying greater prestige internationally. The war had reduced Pakistan's ability to pose a military threat to India and exposed the limitations of the Americans and the Chinese to intervene in an Indo-Pak conflict. In such circumstances, the only reason for the decision could be to consolidate India's dominance and prestige in the region. Similarly, Rajiv Gandhi, who on the one hand gave a plan for the elimination of nuclear weapons at the UN in 1988–89, gave the go-ahead signal to DRDO and the IAEC for the development of an Indian nuclear weapons capability on the other.[120] Inder Kumar Gujral, who was seen as a pacifist, contemplated the possibility of carrying out a nuclear test to enhance his political stock.[121] In addition to the abandoned effort by the Narasimha Rao government in 1995, Vajpayee in his rather short stint of fifteen days in office in the spring of 1996 had ordered Abdul Kalam and Chidambaram to go ahead with the nuclear tests. The scientists initiated the preparations and even placed a device in a test shaft. However, due to his failure to win a vote of confidence from the parliament, the tests could not be conducted.[122]

Public opinion surveys conducted in India after the 1998 nuclear tests clearly showed that 'pride' and 'self-esteem' were the most common themes. In one such poll, conducted by the Indian Bureau of Market Research, as much as 91 per cent of those surveyed felt proud over India becoming a nuclear power. Addressing the Indian parliament, Prime Minister Vajpayee pronounced that, 'It is India's due, the right of one-sixth of humanity.' In a 1994 poll, 49 per cent of those favouring weaponisation of the nuclear option advocated the development of nuclear weapons by India to 'improve its bargaining position in international affairs', while another 38 per cent favoured the acquisition of nuclear weapons to boost India's international stature. Opinion surveys in the wake of the 1974 and 1998 nuclear tests indicated that almost 90 per cent respondents took pride in India's achievement and thought it would enhance its status.[123]

Status consciousness is also apparent in the Indian leadership's public pronouncements. In an interview with the American National Public Radio (NPR), then Foreign Minister Jaswant Singh claimed that, 'All that we have done is given ourselves a degree of strategic autonomy by acquiring those symbols of power ... which have universal currency.'[124]

The theme of nuclear weapons as a means to gain prestige and power keeps appearing regularly and is demonstrated vividly in the following statement by I. K. Gujral, wherein he said that:

> An old Indian saying holds that Indians have a third eye. I told President Clinton that when my third eye looks at the door into the Security Council Chamber it sees a little sign that says, 'only those with economic wealth or nuclear weapons allowed.' I said to him, 'it is very difficult to achieve economic wealth.' The implication was clear; nuclear weapons were relatively easy to build and detonate and could offer an apparent shortcut to great power status.[125]

In a post-nuclear tests analysis, the CIA determined that after a demonstration of its nuclear prowess, India believes that it will now be taken seriously by the world community and its long cherished dream of gaining a permanent seat at the UNSC would also be fulfilled besides bestowing it a privileged place in the regional forums as well.[126] Over the last two decades, India's international stock has definitely increased as is evident from the country specific Civil Nuclear Agreement with the US, the exceptional waiver by the NSG, its entry into the MTCR, the Wassenaar Arrangement, and the Australia Group, besides its growing strategic affinity with the US, though its efforts to gain a permanent presence at the UNSC has so far been frustrated.

India announced a draft nuclear doctrine on 17 August 1999 and a brief official document on 4 January 2003. It has established a three-tiered command and control structure and a tri-service strategic forces command. In 2005, it promulgated its nuclear export control law. However, it has not yet been able to establish an autonomous regulatory authority and sufficient details about its nuclear regulatory and safety and security regimes are not available.

An Analysis of the US–India Nuclear Deal and India's Future Nuclear Ambitions

After having experienced difficulties in pursuit of indigenisation for decades, India finally entered into a civilian nuclear cooperation agreement with the US in 2006. This was followed by the 123 Agreement between the two countries and a special waiver by the NSG in 2008. It is clear that India is now willing to accept safeguards on imported light-water reactors as on indigenous reactors using imported fuel. With most of its reactors likely to run on imported fuel, its domestic uranium reserves would now be available for fissile material production for bomb purposes. India, over a long period, has also accumulated large stocks of reactor-grade plutonium-laden fuel, some of which has already been reprocessed. Under the terms of the Indo-US agreement, India placed eight of its power reactors under IAEA safeguards. However, the agreement failed to bring under safeguards huge amounts of spent fuel already generated by these reactors. According to a report prepared by the International Panel on Fissile Materials (IPFM) in 2015 titled 'Plutonium Separation in Nuclear Power Programs', India has 4,100 to 5,200 tons of spent fuel from various pressurised heavy-water reactors (PHWRs), out of which 2,500 to 3,600 tons is unsafeguarded and once reprocessed, can produce 11 to 13.5 tons of separated plutonium. This is far in excess of India's perceived requirement of its planned fast breeder reactors and experimental MOX fuel assemblies.[127]

India had declared its intent to install 10,000 MW of nuclear energy generation capacity by the year 2000. If this ambitious target had been achieved, nuclear energy would have constituted 5 per cent of India's total energy production. However, there was a lot of scepticism about the IAEC's claims.[128] The performance of the Indian nuclear establishment and implementation of its various programs also came under serious criticism as these were seen as an effort to juxtapose an advanced technology over India's primitive industrial base. It was felt that due to a lack of meticulous planning and absence of any public debate, there were frequent revisions in the plans later on. As a result, progress was unsatisfactory, which led to increasing dependence on foreign assistance and acceptance of safeguards in spite of the self-professed goal of indigenisation and self-reliance.[129] This criticism was proven to be correct as the share of nuclear energy in the overall energy mix remained less than 3 per cent with an installed capacity of 3,750 MW. Indian plans are aimed at achieving 20,000 MW of nuclear energy by 2020 and by 2050, nuclear power is projected to constitute 20 per cent of the total power production.[130] As of May 2017, India's installed nuclear capacity amounted to 6,780 MW and ten nuclear power plants with a capacity of 7,000 MW are planned to be constructed besides 4,300 MW capacity power plants under construction. But it is not clear when these reactors would come on line. It is apparent that the goal of production of 20,000 MW of nuclear energy will not be met as has been traditionally the case with IAEC's pronounced targets in the past, despite the fact that India had signed the Civil Nuclear Agreement with the US and had been granted a special waiver by the NSG a decade ago.[131]

Notes and References

1. 'India Conducts 3 Nuclear Tests', *The Financial Express*, 12 May 1998; Shekhar Gupta, 'Road to Resurgence', *The Indian Express*, 12 May 1998.

2. *SIPRI Yearbook 1975* (Cambridge Massachusetts and London: The MIT Press), p. 16.

3. Ibid., p. 17; K Subrahmanyan, 'India's Nuclear Policy', in Onkar Marwah and Ann Schulz (eds.), *Nuclear Proliferation and Near Nuclear Countries* (Cambridge Massachusetts: Ballinger Publishing Company, 1975), p. 141; Onkar Marwah, 'India's Nuclear and Space Programmes: Intent and Policy', *International Security*, vol. 2 (Fall 1977), p. 98.

4. Subrahmanyam, p. 141; also see *SIPRI Yearbook 1975*, p. 17.

5. Marwah, p. 98.

6. Ibid.

7. Andrew Koch, *Selected Indian Nuclear Facilities* (Monterey, California: Center for Non-proliferation Studies, Monterey Institute of International Studies, 1999), p. 3; Central Intelligence Agency, Scientific Intelligence Report, 'Indian Nuclear Energy Program', 6 November 1964;

'US Intelligence and the Bomb', *National Security Archive Electronic Briefing Book No. 187*, Washington DC.

8. Ibid.

9. Prime Minister's suo moto statement during discussion on 'Civil Nuclear Cooperation with the US: Implementation of India's Separation Plan', *The Hindu*, New Delhi, 7 March 2006.

10. *Scientific Intelligence Report*, Indian Nuclear Energy Programme, op. cit.; Andrew Koch, *Selected Indian Nuclear Facilities*, op. cit., p. 3.

11. *SIPRI Yearbook 1975*, p. 17. Andrew Koch, op. cit.

12. CIA, *Scientific Intelligence Report*, 'Indian Nuclear Energy Program', 26 March 1958; *National Security Archive Electronic Briefing Book No. 187*, Washington DC.

13. Department of State, Telegram to American Embassy New Delhi and American Embassy, London, 24 May 1966; Department of State, Telegram to American Embassy New Delhi with information to American Embassy, London, American Consulate, Bombay, and American Embassy, Rawalpindi, 28 July 1966.

14. Department of State, Airgram No. A-253, 29 April 1964 from American Consulate, Bombay, to the Department of State, 'Inauguration of Indian Plutonium Separation Plant',

15. *SIPRI*, p. 18. Also see Marwah, p. 101.

16. CIA, Scientific Intelligence Report, 'Indian Nuclear Energy Programme', 6 November 1966 and CIA, Directorate of Science and Technology, Scientific Intelligence Digest, December 1965.

17. Department of State Telegram No. 15545, 27 February 1965, 'For Governor Harriman from the Secretary'.

18. Memorandum for the President, NSC Meeting, 9 June 1966, prepared by Acting Secretary of State George Ball, 7 June 1966, p. 4.

19. CIA, Intelligence Information Cable, No. 99143, 24 October 1964.

20. Ibid.

21. Ibid.

22. Special National Intelligence Estimate Number 31-1-65, 'India's Nuclear Weapons Policy', Directorate of Central Intelligence, 21 October 1965.

23. Ibid.

24. Ibid.

25. Ibid.

26. Scientific Intelligence Digest, December 65, op. cit.

27. Ibid.

28. Department of State Telegram No. 26616, 28 July 1966, addressed to American Embassy, New Delhi.

29. Department of State Airgram, No. A–256, to American Embassy, New Delhi, 'Possible Indian Nuclear Weapons Development', 29 March 1966.

30. Memorandum for the President, 9 June 1966, op. cit.

31. Department of State, Telegram No. 13982 of 24 May 1966, addressed to American Embassy, New Delhi, with information to American Embassy, London.

32. Ibid.

33. CIA, Office of Scientific Intelligence, Weekly Surveyor 20 June 1966.

34. Ibid.

35. Memorandum for the President, op. cit.

36. Ibid.

37. Ibid.

38. Ibid.

39. Ibid.

40. Memorandum for Secretary of Defence—JCSM-2-67 of 4 January 1967, 'The Indian Nuclear Weapons Problem: Security Aspects'.

41. Ibid.

42. Memorandum of Conversation between President Johnson and L. K. Jha on 19 April 1967.

43. Department of State, Memorandum of Conversation between Secretary of State Dean Rusk and Soviet Foreign Minister Gromyko, 23 June 1967.

44. US Government Aide Memoire, presented to the IAEC in Bombay on 16 November 1970.

45. Ibid.

46. Ibid.

47. National Security Agency, Communication Intelligence Report No. 3/00/QOC/R668-72 of 31 August 1972. [NSA/DTA NSA/HCF-ZPCA].

48. SIPRI, p. 16.

49. Ibid.

50. US government 'Aide-Memoir' to IAEC, op. cit.

51. Marwah, p. 104.

52. Subrahmanyam, op. cit., pp. 141–2.

53. Rodney W. Jones, 'India', in Jozef Goldblat (ed.), Non-Proliferation: The Why and the Wherefore (London and Philadelphia: Taylor and Francis, 1985), p. 102.

54. William Burr, 'A Scheme of "Control": The United States and the Origins of the Nuclear Suppliers' Group, 1974–1976,' International History Review, April 2014. Available at <http://www.tandfonline.com/doi/full/10.1080/07075332.2013.864690>

55. 'Assessment of Indian Nuclear Test', US mission to NATO, 5 June 1974.

56. Ibid.

57. SIPRI, p. 19.

58. Ibid., p. 20.

59. Ibid., p. 21.

60. CIA, Central Intelligence Bulletin, 20 May 1974.

61. SIPRI, pp. 21–2.

62. Subrahmanyam, pp. 138–9.

63. Ibid., p. 143.

64. Dept. of State, Intelligence Note, 13 June 1974.

65. Ibid.

66. Ibid.

67. Robert Jackson, South Asian Crisis (London: Chatto & Windus, 1975), p. 226.

68. US Department of State Telegram No. 15545, to American Embassy at Tel Aviv with information to

Embassies at New Delhi, Moscow, Hong Kong, and Karachi, from Secretary of State Governor Harriman.

69. Jones, p. 113.
70. Ibid.
71. CIA, *Central Intelligence Bulletin*, 20 May 1974.
72. 'Post Mortem Report—An Examination of the Intelligence Community's Performance Before the Indian Nuclear Test of May 1974', for the Director Central Intelligence, July 1974.
73. Amitabh Mattoo, 'India's Nuclear Policy in an Anarchic World', in Amitabh Mattoo (ed.), *India's Nuclear Deterrent—Pokhran-II and Beyond* (New Delhi: Har Anand Publications, 1999), p. 18.
74. Dhurva Research Reactor, Nuclear Threat Initiative <http://www.nti. org/learn/facilities/837/>
75. *Indian Defence Review Digest*, vol. 4 (New Delhi: Lancer Publications, 1992).
76. Marwah, p. 103.
77. Jones, in Jozef Goldblat (ed.), op. cit., p. 114.
78. Bhabhani Sen Gupta, *Nuclear Weapons? Policy Options for India*, Sage Publications, New Delhi, Beverly Hills, London, 1983, p. 9.
79. Ibid., p. 10.
80. Andrew Koch, op. cit., p. 3.
81. Ibid.
82. CIA, Directorate of Intelligence, 'India's Nuclear Programme—Energy and Weapons: An Intelligence Assessment', July 1982.

83. CIA, Directorate of Intelligence; 'India's Potential to Build a Nuclear Weapon—An Intelligence Assessment', July 1988.
84. Report from US DAO, New Delhi, to JCS, DIA, CIA, 'An Update on Government of India Nuclear Programme', November 1986.
85. Ibid.
86. 'India to get nuclear fuels from Russia', *Dawn*, Karachi, 15 March 2006.
87. Mattoo, op. cit., p. 18.
88. 'Hide 'n Seek: CIA Muse over N-Tests', Associated Press, *The Indian Express*, 18 May 1998.
89. BJP Election Manifesto 1998, pdf - BJP e-Library <lib. bjplibrary.org/jspui/.../241/1/ BJP%20ELECTION%20 MANIFESTO%201998.pdf>
90. Ibid.
91. Manvendra Singh, 'Showcase of technological leap by Indian Nuclear Establishment', *The Indian Express*, 12 May 1998.
92. Ibid.
93. 'Scientist raises doubt about success of Indian N-test', *Dawn*, 28 August 2009 <https://www.dawn.com/ news/849091/scientist-raises-doubt-about-success-of-indian-n-test>; 'Pokhran-II thermonuclear test a failure', *The Hindu*, 17 September 2009, updated 17 December 2016 <http://www. thehindu.com/opinion/op-ed/ Pokhran-II-thermonuclear-test-a-failure/article13736892.ece>
94. Manvendra Singh, *Indian Express*, 12 May 1998.

95. Chidanand Rajghatta, 'The Hindu Bomb', *Indian Express*, 21 May 1998.

96. *The Asian Age*, 12 May 1998.

97. Ibid.

98. Ibid.

99. 'Prime Minister Writes to G-8 Heads on Tests', *Times of India*, 14 May 1998.

100. Arati R. Jerath, 'Govt Flashes China Card at the West', *Indian Express*, 14 May 1998.

101. Ibid.

102. Arati R. Jerath, 'Damage Control begins as Govt. Wakes Up to High Cost of China Card', *Indian Express*, 28 May 1998.

103. Intelligence Report, Office of Near Eastern, South Asian and African Analysis, CIA, 29 May 1998.

104. Ibid.

105. Ibid.

106. Ibid.

107. Ibid.

108. Ibid.

109. Chidanand Rajghatta, 'The Hindu Bomb', *Indian Express*, 21 May 1998.

110. Chidanand Rajghatta, 'Patch up with India, US tells Pak', *Indian Express*, 20 May 1998.

111. 'Hide 'n' Seek: CIA muse over n-tests', *The Indian Express*, 18 May 1998.

112. K. Subrahmanyam, *Nuclear Myths and Realities—India's Dilemma* (New Delhi: ABC Publishing House, 1981), pp. vi–vii.

113. Ibid.

114. Gupta, p. 39.

115. Ibid., p. 17.

116. Ibid., p. 17.

117. Ibid., p. 20.

118. Zia Mian, 'Homi Bhabha killed a Crow', in Zia Mian and Ashish Nandy (eds.), *The Nuclear Debate: Ironies and Immoralities,* Colombo: Regional Center for Strategic Studies, July 1998, p. 12.

119. George Perkovich, *India's Nuclear Bomb,* Berkeley, University of California Press, 1999, p. 36.

120. K. Subrahmanyam, '*Indian Nuclear Policy—1964–98*', in Jasjit Singh (ed.), *Nuclear India* (New Delhi: Knowledge World, 1998), p. 44.

121. Amitabh Mattoo, 'India's Nuclear Policy', op. cit., p. 18.

122. Perkovich, op. cit., pp. 374–5.

123. Mattoo, op. cit., pp. 11–15.

124. Perkovich, p. 441.

125. Ibid., p. 400.

126. Intelligence Report, Office of Near Eastern, South Asian, and African Analysis, CIA, 29 May 1998.

127. 'Plutonium Separation in Nuclear Power Programs', IPFM 2015, p. 57.

128. Ibid.

129. Ravindra Tomar, 'The Indian Nuclear Power Programme: Myths and Mirages', *Asian Survey*, vol. 20, no. 5 (May 1980), pp. 529–31.

130. Government of India, Department of Atomic Energy, Annual Report 2005–2006.

131. 'India will build ten new reactors in huge boost to nuclear power', BBC News, 18 May 2017 <http://www.bbc.com/news/world-asia-India-39958299>

2

India's Nuclear Doctrine: Stasis or Dynamism?

Ali Ahmed[1]

India's official nuclear doctrine, adopted by the Cabinet Committee on Security (CCS) on 4 January 2003, has an expansive intent of inflicting unacceptable retaliatory damage of levels characterised in the doctrine as 'massive' to any form of nuclear first use against India or its forces anywhere.[2] The doctrinal tenet in full is: 'Nuclear retaliation to a first strike will be massive and designed to inflict unacceptable damage.'[3] The primary argument for this formulation is that India believes nuclear weapons are a 'political instrument of deterrence.'[4] Admittedly, the intent to go 'massive' in retaliation should ordinarily deter. As against this, the critique has it that the threat of retaliation at massive levels lacks credibility,[5] particularly if the provocation for retaliation is of a lower order using tactical nuclear weapons (TNW). To buttress the critique, the usual illustration is intruding Indian forces being met by a Pakistani TNW attack. In light of Pakistan's higher number of nuclear weapons, it could retaliate in kind. Critics have it that India would be hard put to follow through with its retaliatory intent as declared, as to do so would be disproportionate and escalatory.[6] In light of continuing India–Pakistan hostility, this is no longer an unlikely scenario. The critique, therefore, requires a plausible doctrinal answer.

However, there has been no change in India's official nuclear doctrine over the past decade and a half. This is puzzling, reflecting as it does either a deficit in nuclear learning or that even as India's declaratory nuclear doctrine remains stagnant, overtly and nuclear doctrinal change has been kept outside the open domain. Nuclear ambiguity has overtaken India's nuclear doctrinal glasnost of the turn of the century. In this chapter, nuclear opacity[7] notwithstanding, the finding is that India's official nuclear doctrine of 2003 stands superseded by doctrinal innovation. There is apparently a move towards a war fighting nuclear doctrine and posture, albeit one to cover the deterrence deficit identified. Since secrecy attends India's operational doctrine,[8] a definitive answer is precluded. This is reason enough to interrogate India's nuclear doctrine.

The chapter tries to answer the following questions: Has there been a change in India's nuclear doctrine? If not, why not? If it has changed, why has not India changed its declaratory nuclear doctrine? The chapter is laid out in two parts. The first part looks through a 'levels of analysis' framework. Kenneth Waltz posited the 'levels of analyses' in his doctoral thesis on the causes of war, published as *Man, the State and War*.[9] He delineated three 'images': systemic or structural, dealing with the distribution of power among states; the nation-state or unit level; and lastly, that of the individual or decision maker. Following Graham Allison, this chapter posits an additional bureaucratic level between the unit and individual levels.

In the first part, a mini-case study on India's Cold Start doctrine— the army's conventional war doctrine predicated on limited war theory[10]—demonstrates how change can be viewed through the levels of analysis framework. The following sections look at impetus to doctrinal change at the structural, state, and institutional levels. The first deals with the global and regional threat environment; the second covers the developments in political and strategic culture; and the institutional level covers turf wars and parochial organisational interests.

In the second part, the chapter returns to the mini-case study on India's Cold Start doctrine with an intent to see what separates doctrinal innovation from stagnation. The finding is that where drivers are present at all three levels, doctrinal innovation results. Absent driver(s), at any level, doctrinal stagnation results. To look for drivers, the duration since Pokhran II is divided into three periods: the National Democratic Alliance's (NDA) two stints in power (1998–2004 and 2014 onwards) and of the United Alliance (UPA) period (2004–14). In the UPA period, the strategy of restraint at the political level prevented the UPA from acknowledging the change in nuclear doctrine, even as a divergence emerged between the declaratory nuclear doctrine and operational nuclear doctrine. Whereas the NDA II's initial willingness to review the declaratory nuclear doctrine suggested that it could have acknowledged this shift, it has not done so for fear of drawing attention to its nuclear posture at a time when it is getting closer to the United States and Japan. The need to own up to the change is also not critical since it can easily be analytically arrived at that change has indeed taken place and the declaratory doctrine is history.

The chapter arrives at the conclusion that India's declaratory nuclear doctrine is implausible. Acknowledging the shift in owning up to an operational nuclear doctrine at variance with its declaratory doctrine is the need of the hour. Persisting with the declaratory doctrine does nothing to add to Indian security under peacetime conditions when deterrence is in play or in nuclear conflict when nuclear use is contemplated. India would do well to depart from its policy of stasis on nuclear doctrine and acknowledge the dynamism of its operational nuclear doctrine. This would help with firming up nuclear risk prevention and reduction measures, necessary to emplace in peacetime some assurance of escalatory control and de-escalation in conflict, especially since the two sides—India and Pakistan—rightly appear to contemplate limited nuclear war as better than devastating their mutual homeland, the subcontinent.

Part I—From doctrinal transparency to ambiguity

India's official nuclear doctrine adopted by the Cabinet Committee on Security (CCS) in January 2003[11] largely adopted the draft nuclear doctrine put out by the National Security Advisory Board (NSAB) in August 1999 (NSAB 1999). In its summation of India's nuclear doctrine, the CCS press release collated eight doctrinal precepts. These are: credible minimum deterrent; No First Use (NFU); 'massive' retaliation; command and control arrangements; negative security assurance; possibility of a nuclear counter to a major attack by other weapons of mass destruction (chemical and biological); adherence to nuclear non-proliferation tenets; and advocacy for a nuclear weapons-free world. Not all of these doctrinal elements need mention in a doctrine, such as approach to non-proliferation and elimination of nuclear weapons. The key tenets are NFU and the scope of nuclear use, whether higher order attacks or lower order. Higher order is defined here as counter-value and lower order is restricted to counter-military and limited counter-force targeting.[12] If retaliation is restricted to higher order nuclear use, then it is easier to maintain nuclear warhead numbers at a minimum.[13]

Change can be along any of the precepts:

- abandoning minimum in the minimum credible deterrent;
- distancing from NFU;
- shifting from higher order nuclear attacks to also contemplating lower order attacks;
- moving from a reflective command and control status to a high alert one;
- withdrawal of or caveating negative security guarantees;
- excluding non-nuclear weapons of mass destruction as triggers for nuclear retaliation;
- leaving out nuclear policy tenets from doctrine, such as approach to nuclear non-proliferation and elimination of nuclear weapons.

Though there are also other directions along which India's nuclear doctrine is potentially liable to change, specifically the NFU[14] and arsenal size (minimal), this chapter limits its discussion to the levels of nuclear retaliation. These are currently pegged at 'massive' or higher order levels of nuclear attack. India believes nuclear weapons are political weapons and not for military use. Its nuclear doctrine is thus based on higher order nuclear retaliation, usually depicted as city busting. This puts a premium on the deterrence function of its nuclear arsenal. It retains the formulation of 'massive' retaliatory intent in its nuclear doctrine.

The Levels of Analysis Framework

An Illustrative Case: Cold Start

Doctrine development, and in turn doctrinal change, is not confined to the military sphere. Doctrine is the foundation of decisions ranging from acquisition of assets to deployment and employment of military resources. Consequently, to view impulses to doctrinal change, a look through the levels of the analysis framework bears recompense. An illustration is in order.

The Indian army's conventional war doctrine underwent a shift over this century. The conventional war doctrine dates to the inability of India's strike corps to rise to the occasion when confronted with mobilisation in wake of the terror attack on India's parliament in December 2001.[15] Their slowness off-the-blocks led to rethinking on readiness and an offensive addition to the conventional punch. The shift is better known colloquially as 'Cold Start', otherwise called 'proactive operations' doctrine. After a long while, ownership of the doctrine was admitted to by the Indian army chief, General Bipin Rawat, early in his tenure.[16]

In the earlier land warfare doctrine, the role of strike corps was one of undertaking countervailing deployments and counteroffensives. This was predicated on the Pakistanis being proactive in taking to the offensive as part of their doctrine of offensive defence.[17] However, after the terror attack on parliament, India chose to be proactive at the sub-conventional level. The shift itself involved a transformation of 'holding' corps in a defensive, border-holding, role to 'pivot' corps, with offensive assets. Additionally, the strike corps would deploy 'integrated battle groups' to launch operations in an earlier timeframe. These are to create conditions for strike corps coup-de-main operations.[18] This shift can be explained by a look at the levels of the analysis framework.[19]

Over the 1990s, at the structural level, India's threat perception was complicated by Pakistan's alleged resort to proxy war. This was attributed to an ossification in India's conventional deterrence. India, therefore, needed to refurbish its conventional deterrence.[20] This it proceeded to do by building a readiness in its offensive formations for conventional retaliation in real time. At the next level is the 'unit' level. The principal movement at this level has been, in India's politics, the shift towards the right side of the political spectrum brought about by the rise of cultural nationalism, espoused by India's leading conservative party, the Bharatiya Janata Party (BJP). This has led to the development of a more assertive strategic culture.[21] This contrasts with the earlier question of whether India had a strategic culture at all.[22] The nuclear tests at Pokhran in May 1998 are an example. The shift to Cold Start is an outcome of the hardening in strategic culture.

Peering within the 'box', at the institutional level is the military itself. The army had to come up with a doctrinal product of its own in the face of doctrinal effervescence in the other two services. The air force got off the doctrinal blocks early with its doctrine in the mid-1990s. The navy was also readying to launch a maritime doctrine. Affected by discussion on continuing utility of military force in the nuclear age, the army quickly staked out its role in conflict. Within the army, the extensive Kashmir engagement had brought to fore the infantry lobby riding on the counter insurgency commitment. The mechanised forces used the opportunity of Operation Parakram—the mobilisation in the wake of the attack on the Indian parliament—to break out of their relegation by reworking the conventional doctrine.

Peering through the Framework

This brief survey of change in the Indian army's conventional doctrine indicates scope for a similar analysis in relation to its nuclear doctrine. Upfront, the nuclear doctrine has not changed. However, the doctrinal debate has grappled with the challenge posed by Pakistani acquisition and advertisement of a TNW capability.[23] It is inconceivable that the change and the debate has not influenced nuclear doctrine. This section attempts to discern if and to what extent the nuclear doctrine has been affected.

The Structural Level

When India went overtly nuclear in 1998, it had cited the nuclear asymmetry with China and Chinese nexus with Pakistan as informing its decision. It has attempted to link its nuclear capability to the threat from China[24] as part of its de-hyphenation from Pakistan. However, adversarial relations with Pakistan ensure that Pakistan cannot be wished away. This is especially so because Pakistan has not abjured from first use of nuclear weapons and India's reliance on conventional retaliation for deterring a sub-conventional threat by Pakistan.

Since both states abide by the NFU, the significance of nuclear weapons in the India–China equation stands considerably receded. The two have a quarter-century-long experience of managing their relationship through confidence building measures and agreements on managing the 4,000 km long border. Their mutual stand down from a potential confrontation at Doklam, on the tri-junction of the India-China-Bhutan border, is a case in point.[25] Not a single soldier has died on either side defending the border nor has a shot been fired either in panic or fear or anger, over the past fifty years since 1967. Their last military confrontation was close to Doklam, at Nathula.

This is not the case with the India–Pakistan relationship. The two have not only fought a border war on the Kargil heights in 1999, but have been engaged in sub-conventional operations along the Line of Control (LoC), within Kashmir, and, if Pakistani allegations are taken into account, in proxy wars within Afghanistan and in Pakistan. India stepped up its sub-conventional operations with the launch of trans-LoC 'surgical strikes', responding to a terror attack in the Uri sector in Jammu and Kashmir.[26] For its part, Pakistan denied the raids took place, reducing the onus on it to retaliate.[27] The rocking of the unwritten ceasefire agreement along the LoC that dates to 2003 is indicative of the escalatory dangers. These have been subsumed lately in India's 'two-front war' formulation, a conflict in which India sees itself hemmed in by its two collusive adversaries.[28]

The usual scenario has a trajectory of events, sparked off by a terror attack. India is depicted as launching Cold Start. Pakistan, fearing worse could follow, resorts to TNW(s). India's nuclear doctrine posits massive nuclear retaliation. Pakistan, having taken care to build up its nuclear arsenal into three digits, appears to have enough nuclear ordnance to hurl back. To Khalid Kidwai, a former head of Pakistan's Strategic Plans Division, this implies that South Asia is in its era of mutual assured destruction (MAD).[29] The declaratory doctrine does not provide an adequate answer to this critique.[30] Some strategists from the military school argue that this can be done by continuing with delivering unacceptable damage.[31] Others have argued that going in for unacceptable damage may lead to self-deterrence in a situation of MAD.[32]

One answer, attempted by the head of the National Security Advisory Board in 2013, was in insisting on the declaratory doctrine, arguing that there is no guarantee that escalation would not reach such levels eventually.[33] A former chief of India's Strategic Forces Command

(SFC) also favours status quo.[34] In a recent book of reminiscences, former National Security Adviser Shiv Shankar Menon informs that India contemplated resort to first strike levels of attack in case of Pakistani resort to TNW, and even when readying to do so in a preemptive mode.[35] This suggests two lines of Indian response: status quo and change. There is a third possibility, of change in terms of proportionate response.[36] A TNW attack could be met with requisite levels of retaliation, quid pro quo, or quid pro quo plus.[37]

Of the three options, the first two have an underside. Retaliating massively or with first strike imply a large arsenal in order to ensure against a similar counterstrike by Pakistan. The number of Indian warheads are marginally less than those of Pakistan.[38] India is not markedly ahead of Pakistan in delivery capability in terms of missiles. It has advantages in air delivery and is working on a missile defence shield of dubitable efficacy.[39] Thus, even first strike levels of nuclear ordnance cannot prevent a Pakistani counter of unacceptable proportions, enough to invoke self-deterrence in regard to higher order nuclear retaliation. This would not be the case in relation to the third option: proportionate response. Having detonated three sub-kiloton warheads of the five explosions during Pokhran II, India has the warheads and, in its air vectors, the delivery capability. The option preserves against default escalation and, on that count, is credible.

India is hard put to acknowledge this logic since it provides Pakistan a window to go for nuclear first use with an assurance against escalation, debilitating Indian nuclear deterrence. The nuclear cloud puts paid to India's conventional operations, depriving India of its conventional advantage. India would also be unable to deter Pakistan at the sub-conventional level. Thus, Indian deterrence would suffer across the three levels of conflict: nuclear, conventional, and sub-conventional. India requires denying Pakistan any sense of impunity. At the structural level, the unwillingness to incentivise Pakistani TNW use accounts for India's reticence in acknowledging doctrinal change.

The structural explanation intermeshes with the political level (more in the following section). Besides, induction of TNW into the arsenal negates the 'minimum' tenet.[40] Since it has two nuclear-armed adversaries to cater for, one reckoning has it that India needs 600 strategic nuclear weapons and 300 TNWs.[41] The elasticity is permitted by India's argument that weapon numbers are contingent on those of its two adversaries. Paying lip service to minimum is required for political reasons to depict India as a responsible nuclear power not engaging in an arms race. Therefore, even if India has the wherewithal for proportionate response for structural reasons, it cannot own up to this for political reasons at the next level.

The Political Level

Over the past thirty years, the primary political shift in India has been towards the right side of the political spectrum. The conservative party, the BJP, representing cultural nationalism,[42] strode onto the national stage, beginning with the demolition of the Babri Masjid in the early 1990s. It has since then led the coalition in power between 1998 and 2004 and is currently the ruling party with a majority in parliament, a first for some three decades. In its first stint in power in 1996, that lasted 13 days, it had authorised the nuclear tests. When in power next at the head of the NDA coalition, it conducted the nuclear tests in May 1998. The nuclear doctrine had NDA imprimatur.[43] It was also in the BJP-led NDA term in 1998–2004 that the Cold Start doctrine was conceptualised, though released early in the tenure of its successor UPA government. Other influences such as steady economic growth, increasing political salience of the middle class, and influence of the influential diaspora have also been incidental.

These factors have affected India's strategic culture. Under their cumulative influence, Indian strategic culture has shifted from being defensive and reactive to self-regarding assertion. India's national power indices have grown in the period. Its military investment has steadily increased, with its budget expanding on higher growth rates. Correspondingly, India's strategic doctrine—approach to the use of force—has shifted towards the offensive segment, with strategic doctrine imagined as located along an appeasing-defensive-deterrent, offensive-compellence continuum.

The term 'massive' in the doctrine was a step ahead of the formulation in the draft nuclear doctrine that restricted such retaliation to being of sufficient levels to cause unacceptable damage.[44] The official doctrine, adopted in January 2003 in the immediate wake of the ending of Operation Parakram, incorporated the term for political messaging directed at Pakistan. An Indian analyst suggests its political origin and political utility,[45] implying that it is not quite a strategic yardstick.

With a self-image of a regional power and an aspiring great power, India does not prefer to be boxed in with Pakistan into South Asian confines. Pakistan's going in for TNWs implies asymmetric escalation.[46] This is to extend nuclear deterrence to cover the conventional level, as part of 'full spectrum deterrence'.[47] India, in rebound, projects disdain, relying instead on escalatory nuclear retaliation. It has simultaneously taken care to configure its conventional doctrine to be limited war compliant, cognizant of possible nuclear thresholds. This would put the onus of nuclear escalation onto Pakistan, seemingly legitimising an Indian counterstrike of an overwhelming order. India believes it would survive owing to its size, governmental reach, and national coherence. Pakistan lacks strategic depth and loss of a few major cities could set it back inordinately.[48]

India hopes to release itself from Pakistani shackles but disregards the vulnerability of India's economic rise to counter retaliation. India gambles on its evolving missile shield, at a minimum to cover its economic and political hub centres, Mumbai and Delhi respectively. To cope with a nuclear exchange aftermath, it also has the National Disaster Response Force, a vast central police and a paramilitary reserve, to include the army's Rashtriya Rifles and Assam Rifles. This provides the reassurance to India to continue with an unchanged nuclear doctrine, in seeming disregard to the advent of MAD.

As regards China, India has over the past decade been conventionally minded, creating the border infrastructure and raising mountain divisions. Two mountain divisions in a defensive role and one mountain strike corps of two divisions and armoured brigades[49] are in place. It is not reliant on nuclear weapons. But it needs to stare down China's Rocket Forces, elements of which are deployed close to India's heartland, on the Tibetan plateau. India is ensuring a second strike capability based on nuclear-armed submarines, long-range missiles, multiple independently targetable re-entry vehicles, and thermonuclear warheads.[50] In the context of India's nuclear direction of holding Chinese cities on its eastern seaboard hostage, 'massive' is not inapt. It has political dividend in enabling India to play in its preferred league, equated with China.

Though the nuclear doctrine dates to the last NDA stint in government, the continuity in its nuclear doctrine through the UPA period calls for an explanation. The UPA was apprehensive of being outflanked by the BJP on security issues. It felt vulnerable to BJP's charge of being 'soft' on security and—unwilling to pay an electoral price for this—was continually looking over its shoulder on security issues. Under the UPA, the military firmed up the Cold Start doctrine through successive large-scale military exercises. Riding on a high GDP, the defence budget expanded, enabling acquisitions to operationalise the doctrine such as filling of voids in night fighting capability of mechanised forces. Even so, the UPA was unable to follow through with Cold Start when faced with the 26/11 Mumbai terror attacks.[51] The reason was that the makeover was not quite complete.

On the nuclear doctrinal front, the UPA weighed in for declaratory continuity, keeping a lid on the military's thrust for operationalisation. Even so, India's SFC evolved on its watch. Its successive chiefs, on demitting office, have been voluble on the direction of the military deterrent.[52] In its second five years, the UPA II reeled from political setbacks. Though the Pakistanis demonstrated their TNW capability,[53] the government did not undertake a review. The thinking within the National Security Council Secretariat (NSCS) was voiced by Shyam Saran and Shiv Shankar Menon.[54] The Naresh Chandra Task Force deliberations did not cover the nuclear gamut,[55] even though Naresh Chandra, a former bureaucrat, was intimately associated with the development of India's nuclear deterrent. The task force did not go beyond recommending a permanent chairman for the chiefs of staff committee, the military superior of the head of the SFC.[56] Since the National Security Adviser (NSA) is also in the reporting line and is charged with the operational side as head of the executive council of the National Security Council, it is unclear if the military reporting channel of the SFC is only for administrative matters. The last nuclear related initiative of the Manmohan Singh government was a proposal for an international norm on NFU.[57] The internal political angle to this is that it was a last-ditch effort to bolster the NFU under threat from within the Indian strategic community.[58] In the political context of 2014, it was feared that the NDA could return to power and had an agenda to review India's nuclear doctrine, beginning with jettisoning NFU.

As feared, the NDA did return to power, promising a doctrinal change. The BJP manifesto for the 2014 elections referred to the BJP intent to review the nuclear doctrine. In the event, the BJP's prime ministerial candidate Narendra Modi, when on the campaign trail, put a lid on the controversy. He feared it would impact the BJP's developmental plank,[59] opining that the nuclear 'cultural inheritance' would hold. The aim was to reassure prospective economic partners in development, and strategic partners such as Japan and the US. Consequently, India retains its official nuclear doctrine despite developments in the strategic environment and nuclear weapons field that suggest an operational nuclear doctrine could well be different.

The Institutional Level

There are three significant lobbies at play: the scientific enclave, the national security bureaucracy, and the military. Political decision makers are notable for their absence[60]—the hurly burly of Indian politics keeping them away for the most part and their interests seldom including national security. The bureaucrats see their role as balancers and facilitators. It is not known to what extent the prickly relationship between the lobbies has been ironed out with the evolution of the national security bureaucracy.

The scientific lobby is required to furnish the technology for operationalising the doctrine. Between the two possible demands on it—conferring a capability for inflicting unacceptable damage versus enabling war fighting at lower order levels of nuclear exchanges—the scientific lobby may find the former easier to deliver. The latter involves intricacy in command and control, miniaturisation, and multiple delivery means including short range missiles and air power. This is alongside a capability for higher order exchanges. Consequently, the scientific lobby is engaged with every dimension of nuclear weapons related technology.[61] The level of its grasp over each is, however, debatable since it is often criticised for promising more than it delivers.[62] Its concern with 'turf' springs from prospects of empire building, increased budgets, a relative salience in decision making, access to decision makers, and engagement with emerging technologies. Externally, technology demonstrators help politically burnish India's great power aspirations, and internally bring political dividends for technologists in terms of bureaucratic salience. An infamous instance is in its claim that one of the Pokhran II tests was of the hydrogen bomb, revealed later to be a fizzle.[63]

There is also a controller-custodian divergence in the Indian system,[64] in which, while the custodian is the military, the controller is civilian. The technologists would prefer the de-mated format of Indian readiness, with the warheads kept separate from delivery systems. This suggests that the scientific lobby would be in favour of stasis in nuclear doctrine, even as it forges ahead with operationalisation of a variegated nuclear capability. This plausibly accounts for continuity in declaratory doctrine and simultaneously caters for the possibility of a divergence in operational doctrine.

For its part, the military would prefer operationalisation of a doctrine cognizant of the conventional dimension and insists on effectiveness of the weapon systems.[65] Civil-military relations have affected the pace of operationalisation of India's nuclear deterrent by keeping the military out of the loop.[66] The military's conventional role impacts nuclearisation by way of two connotations. One is to thrust upwards the low nuclear threshold projected by Pakistan so as to create the space for conventional retribution. Secondly, the military is pushing for operationalisation of the nuclear deterrent with a view to provide a variety of options to the decision makers.[67] Scope for declaratory doctrine being distinct from operational doctrine exists. Absence of change in the declaratory doctrine does not imply operational status quo.

Bureaucratic politics can be expected in the national security establishment vested with nuclear thinking. The pulls and pressures of institutions in the nuclear field require arbitration, enabling bureaucratic intervention, if not usurpation, of the field. Bureaucrats self-selecting to be national security minders have carved out a significant role, intervening between the technologists and the military on the one hand and the political level on the other. Their mastery of governmental processes, in particular financing of projects, accounts for their power. The influence of the bureaucrats is non-trivial. An illustration of their being at the apex of the policy loop is the case of the national security adviser cum principal secretary, Brajesh Mishra, attempting to adjudicate on the controversy within the scientific community on whether the hydrogen bomb test was a dud.[68] K. Subrahmanyam, a bureaucrat, who had self-selected to study national security issues in the 1960s and the nuclear aspect in particular, was the head of the NSAB, which drafted the 1999 nuclear doctrine.

The national security bureaucrats then at the helm of the NSCS included the controversial term 'massive'. Since a change would imply reversion to the 1999 formulation and amount to an acknowledgement of a gaffe, there is an incentive to stick with the phrase. Also, most of the people working on the nuclear doctrine have been influenced by Subrahmanyam, since he was the long standing doyen of the strategic community. Subrahmanyam was for counter-value targeting in order to ensure that numbers remained at a minimum and to accentuate the political aspect. The inflection 'massive' appears to be overenthusiasm meant for overawing Pakistan.

Such errors recur too often for comfort that deterrence is in safe hands.[69] 'Massive' retaliation is promised in response to 'first strike'. The drafters apparently meant 'first use', not 'first strike'.[70] 'Massive' is itself a misapplied lift from the US doctrine of early cold war vintage, massive retaliation. While India wishes others to believe its NFU pledge, in the draft nuclear doctrine, NFU was qualified by, and confused with, the negative security guarantee.[71] NFU was not taken as applicable for states aligned to the state that resorted to nuclear first use. This phrasing was deleted from the 2003 official doctrine.

Nevertheless, confusion between NFU and negative security guarantees continued. The NFU is a guarantee to states with nuclear capability and the ability to resort to nuclear first use should they so choose. On the other hand, negative security guarantees are applicable to non-nuclear states. To the NFU policy, the official doctrine quite unnecessarily added: 'Non-use of nuclear weapons against non-nuclear states.' Further, it ended up introducing another caveat to its NFU pledge, this time over major attack by the other two weapons of mass destruction. At the golden jubilee of the National Defence College in 2010, the otherwise erudite NSA, Shiv Shankar Menon, made a gratuitous error in referring to NFU as, 'no first use against non-nuclear weapon states.'[72]

A self-awareness perhaps accounts for lack of bureaucratic confidence in taking doctrinal thinking further. This explains why the doctrinal field at the conventional level is left wholly to the military. The joint doctrine of the armed forces, released in April 2017, also appears to be a unilateral work of the military.[73] By staying out of what they consider military turf at the conventional level, the bureaucrats have appropriated the nuclear level and kept the military out. This is reinforced by the well-known antipathy between the bureaucrats and the brass.[74] The bureaucrats would unlikely weigh in easily in favour of operationalisation that gives the military control and, with it, greater institutional power.

Even in case of operationalisation, the chain of command continues to vest with the bureaucrats. Militaries report to elected civilian heads, not to bureaucrats. The current command and control arrangement has the military head of the SFC reporting to both the NSA and the Chairman, Chiefs of Staff Committee (COSC). The latter is double-hatted as the head of his service and, privileging his service, ordinarily has little time for his additional role.[75] There is no support staff within the Headquarters Integrated Defence Staff—which serves the COSC—having a nuclear decision-making support role. The headquarters SFC cannot substitute for this since it would amount to writing out their own orders. In contrast, the NSA has a strategic and defence division in the NSCS for support. Thus, an inexplicable situation arises in which the uniformed SFC ends up taking his primary cue not from his uniformed superior, the Chairman COSC, but from the NSA, a bureaucrat with his tenure co-extensive with that of the prime minister who appoints him.[76]

Any doctrinal shift towards proportionate response would involve a greater military engagement with the conflict strategy and, in turn, nuclear strategy. This would entail reworking the command and control arrangements for greater inclusivity of the military.[77] The war strategy, military strategy, and nuclear strategy need to be intertwined. This perhaps accounts for a command and control arrangement unsustainable in organisational theory and military culture. Declaratory stasis, therefore, overshadows the dynamism in operational nuclear doctrine.

Part II—Innovation Versus Stagnation

Explaining Doctrinal Change

The illustration of Cold Start in the last part divulges that factors for change (or status quo) in India's nuclear doctrine can be located at all three levels. It can be said that the primary impetus is difficult to discern. Instead, factors at all three levels appear complementary and responsible for doctrinal change or otherwise. To find out the relative salience of the factors, a reversion to the mini-case study on Cold Start is in order.

Cold Start and Drivers of Doctrinal Change

The antecedents of the Cold Start doctrine can be traced to the 1971 war. The period since can be divided as follows. First is the period of operation of the Indira Doctrine that includes the regimes of Indira Gandhi and Rajiv Gandhi; second is the coalition governments of the 1990s; third is the onset of cultural nationalism in the NDA I period; fourth, the UPA decade; and finally, the ongoing NDA II period. The periods are contrasted in respect of innovation and vice-versa in doctrinal thinking in the conventional sphere.

The period of the Indira Doctrine—India's Monroe Doctrine—was one of doctrinal innovation. The army shifted to mechanisation. The second period, covering the 1990s, witnessed India being bogged down by counter insurgency operations, resulting in ossification of its conventional deterrent based on unwieldy strike corps. The following period (NDA I) was one of doctrinal innovation with India forging a limited war doctrine. In the UPA period, doctrinal dissonance prevailed since the military continued down the Cold Start path, putting it at odds with the doctrine of strategic restraint. The NDA II period is one of doctrinal innovation, with the army taking public ownership of Cold Start and introducing aggression such as trans-LoC surgical strikes.

In periods of doctrinal innovation, it is seen that factors at the three levels—structural, political, and organisational—are in operation. In the Indira–Rajiv period, at the structural level, the twinned nuclear and sub-conventional threat from Pakistan led to India leveraging its conventional advantage through a shift to mechanisation. The Indira Doctrine shifted Indian strategic culture away from Nehruvianism of the preceding era. The Pokhran I Nuclear Test, viewing the Indian Ocean as an Indian lake, and intervention in Sri Lanka, etc. are expressions, and evidence, of this. Organisational impetus to doctrinal change was provided by the dynamism injected by the likes of the thinking general, K. Sundarji.[78]

Likewise, in the period of innovation under the NDA I, drivers at the three levels can be seen at work and in sync. At the structural level, the threat from Pakistan peaked in the Kargil War and terrorist provocation under nuclear conditions thereafter. At the political level, strategic culture was impacted by changes in politics brought on by mainstreaming of cultural nationalism. At the organisational level, coping with the new nuclear reality and stepped-up inter-Services competition over changes in the character of war brought on by the revolution in military affairs led to articulation of doctrines by all three services.

Conversely, doctrinal stagnation is a product of lack of impetus at any level—structural, political, or organisational. For instance, if there is lack of threat, there is little incentive to innovate. At the political level, the strategic culture must be receptive to change. This may not be so in case of weak polities. At the organisational level, stagnation can result from complacency, lack of budgets, concentration in non-core areas, and absence of doctrinal entrepreneurs.

In the case of India, in the two periods—the coalition era of the 1990s and the UPA decade—there was a lack of a driver at some levels. In the 1990s, the political level exhibited a deficit in political cohesion, while at the organisational level the army privileged a counter insurgency focus. This led to doctrinal stasis. In the 1990s, forceful articulation of a critique on India's lack of strategic culture was witnessed, giving rise to an aggressive inflection in strategic culture over the succeeding NDA I period. In the second instance of doctrinal stagnation—the UPA decade—the threat at the structural level was dormant. India was engaged in the Composite Dialogue and there was a 'healing touch' operational in Jammu and Kashmir. At the political level, the inflection of cultural nationalism—accounting for aggression in strategic culture—was reined in by a doctrine of strategic restraint. At the organisational level, while Cold Start firmed in, it remained without imprimatur and acknowledgement. The survey above suggests that in case drivers are deficient or absent at any one or more of the three levels, doctrinal stagnation results.

The Nuclear Doctrine: Innovation or Stagnation?

As with Cold Start above, a periodisation can be done briefly to examine the nuclear doctrine beginning with India's nuclear quest dating to the mid-1960s. This line of inquiry is in the tradition of Scott Sagan's work. Scott Sagan identified three 'models' as to why states build or refrain from nuclear weapons: the 'security' model is self-explanatory, and is useful as justification or rationalisation; the 'domestic politics' model, in which nuclear weapons are tools used to advance parochial political and bureaucratic interests; and the 'norms' model, in which pursuit of the nuclear program is an important symbol of a state's modernity and identity.[79] Sagan's first, the security model, is located at the structural level; his second, the domestic politics model, encompasses the political and organisational level, rooted as it is in political culture and bureaucratic politics; and the third, the norms model, is equivalent to the political level.

The Chinese nuclear test in October 1964 concentrated minds and commentary on whether India could and should go nuclear.[80] The next highpoint is in 1974 when in the wake of the Pokhran I Nuclear Test, the debate centred on whether India should go nuclear.[81] The debate continued in the late 1980s, informed by reported developments in relation to Pakistani acquisition of nuclear weapons.[82] The discourse was not only on advisability of going nuclear, but also reflected on a putative nuclear doctrine predicated on a small nuclear arsenal. Naturally, in light of low numbers of warheads and limited technology of delivery, the doctrine was predicated on counter-value targeting. India's political aversion to nuclear weapons and conventional advantages led to NFU acquiring the status as a doctrinal pillar in this incipient nuclear doctrinal thinking.

In the early part of NDA I, Pokhran II compelled the articulation of a doctrine. The Kargil War impelled its articulation by the nuclear doctrine drafting group of the NSAB in 1999 that was adopted by the government in 2003. Though the driver of India's nuclearisation was projected as the long-term threat from China, the key factor at the structural level was the threat from Pakistan. Pakistan seems to have taken recourse to proxy war since the late 1980s ostensibly under cover of its evolving nuclear capability. India's nuclear breakout was to reposition India as an indubitable power with nuclear weapons, shaking it away from its earlier position of non-weaponised 'recessed deterrence'—a term its votary in the period, Jasjit Singh, used to rationalise against going overtly nuclear.[83] The NFU was privileged to project India as a reluctant nuclear adherent and, in light of the NFU, a 'retaliation only' strategy was predicated on higher order targeting. The heightening of the Pakistani threat over the NDA term—which witnessed a near year-long standoff with Pakistan and a fully mobilised military—led to a hardening of the promise of higher order nuclear retaliation reflected in the term 'massive'. Implicit was a preference for a 'minimum' deterrent, reliant on a small nuclear arsenal, in keeping with India's dismissive attitude to cold war nuclear theology and its characteristic, the arms race. The political level witnessed the advent of cultural nationalism into political respectability, impacting in the form of an assertive strategic culture. At the organisational level, the three services were keenly positioning for custody of the nuclear deterrent. The 2003 doctrine empowered the military to raise the SFC and gain joint control over the nuclear assets with the technologists.

The UPA years were of continuity. At the structural level, there was an opening to Pakistan, a legacy of the NDA I period. The UPA was also lobbying for a nuclear deal with the US and wanting to get into the Nuclear Suppliers Group. This necessitated being reticent and not drawing attention to nuclear developments. At the organisational level, there was considerable development in regard to the NSCS and the SFC, and, likewise, on the technological front. The strategic doctrine of restraint was the tool used to police the two. Prodded by India's continuing firming up of the Cold Start doctrine, Pakistan went on to introduce TNW into its arsenal. Though this put India's declaratory doctrine under scan and nuclear operationalisation enabled a shift, the UPA did not budge. Karnad characterises this as 'confusion' and the government being 'seriously conflicted' between idealism and strategic imperatives.[84]

The NDA II period has seen the structural level factor—threat perception—resurface with a dive in India–Pakistan relations. The political level has witnessed the 'Doval Doctrine', named after the NSA, with strategic assertion at its core.[85] At the organisational level, the military has been at the forefront with public articulation of national concerns such as the army chief's 'two-and-a-half front war' formulation.[86] In short, the dynamism in the drivers at the three levels suggests a fertile period for doctrinal innovation. However, the declaratory nuclear doctrine remaining static needs explaining. This can be attributed to Narendra Modi, then prime ministerial candidate, putting a lid on the thrust in the strategic discourse for a change in nuclear doctrine, calling it a 'cultural inheritance'.[87] The thrust line has mainly been on rethinking NFU.[88] The doctrinal reticence regarding arsenal variegation needs an accounting.

The brakes are applied at the political level. India wishes to be in the big league. The global power equations are witnessing a tryst between a hyper-power, the US, and an emerging power, China. India is leaning towards the US, given its proximity with China and the border dispute. Wary of provoking Chinese counter containment, India is, nevertheless, in strategic relationships with balancing powers, the US and Japan. Neither wishes to see India proactive on the nuclear front, doctrinally. India also does not want its nuclear doctrine to be seen as responsive to Pakistani provocations, preferring declaratory continuity to obfuscate over the changed reality. Therefore, India's reticence cannot be read as doctrinal stagnation. The stasis depicted by declaratory doctrine is to serve as a fig leaf for closed-domain doctrinal innovation.

Conclusion

India's tryst with the nuclear doctrine started on the right note. Transparency was the watchword, the deterrence value being in the communication of determination and intent to the adversaries. India was also under pressure to indicate its position on nuclear weapons use after it went nuclear in May 1998. By coming out with a doctrine predicated on NFU and assured retaliation, India hoped to put Pakistan on the back foot. Pakistan wanted to extend its nuclear cover to the conventional level and could not match India's NFU. In the event, Pakistan chose nuclear ambiguity, preferring not to state its doctrine. From its doctrine-related pronouncements and technology developments, Pakistan has a first use doctrine and a low threshold. This has complicated India's declaratory nuclear doctrine in that it cannot credibly continue to project that it would retaliate with higher order strikes to any form of nuclear first use against it. Despite this conundrum, India has chosen to stick with its declaratory doctrine.

Its technological developments suggest that it has other options. This brings to fore the possibility that even if declaratory nuclear doctrine is unchanged, the operational nuclear doctrine diverges.

This chapter has shown that there are drivers for change and also that there are present factors favouring retention of the declaratory doctrine, but only as a cover for such change. At the structural level, India wishes to use its conventional advantage. The best cover for this is not an incredible declaratory doctrine, but an operational doctrine countenancing flexibility. The political level cannot but be cognizant of the MAD situation prevailing in the subcontinent. Consequently, self-deterrence will stay the nuclear hand. At the organisational level, the technological developments have expanded retaliation options. The conclusion here is that India's declaratory doctrine is no longer applicable. India has a different operational doctrine that it must own up to.

Nuclear opacity provides cover but not security. The gains of transparency have been made in projecting India as a responsible nuclear power. It has not forced Pakistan to follow suit with adoption of an overt nuclear doctrine of its own. Ambiguity has a deterrent function in that Pakistan cannot be sure of India's nuclear actions. The constant sniping at NFU within India and indication that India even considered first strike as an option informs of Indian options kept open. Pakistan's nuclear stance, based on TNW, is thus put on notice. It cedes space to India for its Cold Start operations.

However, since a possible terror attack can yet act as a trigger for eventual nuclear exchange(s), the two sides remain insecure. The new nuclear doctrine must acknowledge the MAD circumstance. Security entails doctrinal exchanges and creation of a nuclear conflict management mechanism. The mandate would be to enable mutual escalation control and de-escalation in case of a conflict gone nuclear. The first step towards this is for India to jettison its declaratory doctrine. Its replacement must enable entering into a tacit arrangement with Pakistan that can bring about immediate nuclear exchange termination and early conflict termination.

Notes and References

1. The author is grateful to a former colleague for comments on the draft of this chapter.

2. Prime Minister's Office, 'Cabinet Committee on Security Reviews Progress in Operationalizing India's Nuclear Doctrine', 2003, accessed 8 August 2017 <http://pib.nic.in/archieve/lreleng/lyr2003/rjan2003/04012003/r040120033.html>

3. Ibid.

4. Manpreet Sethi, *Nuclear Strategy: India's March towards Credible Deterrence* (New Delhi: Knowledge World, 2009), p. 126.

5. Sethi, 2009, xxii, p. 125.

6. Ali Ahmed, *India's Doctrine Puzzle: Limiting War in South Asia* (New Delhi: Routledge, 2014), p. 158.

7. Arun Prakash, 'India's Nuclear Deterrent: The more things change…', Policy Report (Singapore: S. Rajaratnam School of International Studies, March 2014), pp. 6–7.

8. Kampani, Gaurav, 'New Delhi's Long Nuclear Journey: How Secrecy and Institutional Roadblocks Delayed India's Weaponization.' *International Security* 38, no. 4 (Spring 2014), pp. 79–114.

9. Kenneth Waltz, *Man, the State and War: A Theoretical Analysis* (New York: Colombia University Press, 1957).

10. Walter Ladwig III, 'A Cold Start for Hot Wars? The Indian Army's new Limited War Doctrine', *International Security*, vol. 32, no. 3, pp. 158–90.

11. Prime Minister's Office, op. cit.

12. Rajesh Rajagopalan and Atul Mishra, *Nuclear South Asia: Keywords and concepts* (New Delhi: Routledge, 2014), pp. 106–7.

13. Shiv Shankar Menon, *Choices: Inside the Making of India's Foreign Policy*, Series: Geopolitics in the 21st Century (Washington DC: Brookings Institution Press, 2016), p. 108.

14. Siddharth Varadarajan, 'Menon: The Policy of No First Use of Nuclear Weapons Has Served India's Purpose', *The Wire*, December 2016, accessed on 15 July 2017 <https://thewire.in/87530/menon-india-nuclear-weapons-nfu-nsa/>

15. Jasjit Singh, 'Doctrine and Strategy Under the Nuclear Overhang', *Financial Express*, 4 June 2004.

16. Sandeep Unnithan, 'We will cross again', *India Today*, 4 January 2017, accessed on 7 August 2017 <http://indiatoday.intoday.in/story/lt-general-bipin-rawat-surgical-strikes-indian-army/1/849662.html>

17. Gurmeet Kanwal, *Indian Army Vision 2020* (New Delhi: Harper Collins 2008), p. 68.

18. Ibid., pp. 85–86.

19. Ali Ahmed, 2014, op. cit., pp. 5–7.

20. Shankar Roychowdhury, *Officially at Peace* (New Delhi: Viking, 2002), pp. 27, 38–9.

21. Tobias Engelmeier, *Nation Building and Foreign Policy in India: An Identity-Strategy Conflict* (New Delhi: Cambridge University Press

India Ltd, 2009), pp. 57–71.

22. Gurmeet Kanwal, *Nuclear Defence: Shaping the Arsenal* (New Delhi: Knowledge World, 2001), pp. 39–41.

23. Peter Lavoy, 'A Conversation with Gen. Khalid Kidwai', Carnegie International Nuclear Policy Conference 2015, 23 March 2015, pp. 4–5 <https://carnegieendowment.org/files/03-230315carnegieKIDWAI.pdf>

24. George Perkovich, *India's Nuclear Bomb: The Impact on Global Proliferation* (Berkeley: University of California Press, 2002), p. 417.

25. Ministry of External Affairs, 'Press Statement on Doklam Disengagement Understanding', 28 August 2017, accessed on 30 August 2017 <http://www.mea.gov.in/press-releases.htm?dtl/28893/Press+Statement+on+Doklam+disengagement+understanding>

26. Press Information Bureau, 'Press statement by DGMO', 29 September 2016, accessed on 3 August 2017 <http://pib.nic.in/newsite/PrintRelease.aspx?relid=151242>

27. Inter-Services Public Relations, 'Press release', 29 September 2016, accessed on 5 August 2017 <https://www.ispr.gov.pk/front/main.asp?o=t-press_release&id=3483&cat=army>

28. Rajat Pandit, 'India cannot rule out possibility of two-front war with China and Pakistan, Army chief General Bipin Rawat says', *The Times of India*, 6 September 2017 <http://timesofindia.indiatimes.com/india/india-cannot-rule-out-possibility-of-two-front-war-with-china-and-pakistan-army-chief-general-bipin-rawat-says/articleshow/60395986.cms>

29. Lavoy, 2015, op. cit., p. 4.

30. P. R. Chari, 'India's Nuclear Doctrine: Stirrings of Change', Carnegie Endowment for International Peace, 4 June 2014, accessed 20 July 2017 <http://carnegieendowment.org/2014/06/04/india-s-nuclear-doctrine-stirrings-of-change-pub-55789>

31. Balraj Nagal, 'India's Nuclear Strategy to Deter: Massive Retaliation to Cause Unacceptable Damage', *CLAWS Journal* (Winter 2015): 1–20, p. 10 <http://www.claws.in/images/journals_doc/440323975_balrajnagal.pdf>

32. Ali Ahmed, 'South Asia: Nuclear Self-Deterrence as a Virtue', *Foreign Policy Journal*, 26 April 2017, accessed on 7 July 2017 <https://www.foreignpolicyjournal.com/2017/04/26/south-asia-nuclear-self-deterrence-as-a-virtue/>

33. Shyam Saran, 'Is India's nuclear deterrent credible?' 24 April 2013, accessed on 25 July 2017 <http://www.armscontrolwonk.com/files/2013/05/Final-Is-Indias-Nuclear-Deterrent-Credible-rev1–2–1–3.pdf>

34. Nagal, 2015, op. cit., p. 14.

35. Menon, 2016, op. cit., p. 117.

36. Ali Ahmed, 'Political decision making and nuclear retaliation', *Strategic Analysis* (New Delhi: Institute for Defence Studies and

Analysis), vol. 36, no. 4, pp. 511–26; July–August 2012, p. 519.

37. Krishnaswami Sundarji, *Vision 2010: A Strategy for the Twenty First Century* (New Delhi: Konark Publication 2003), pp. 146–53.

38. Hans Kirstensen and Robert Norris, 'Worldwide deployments of nuclear weapons, 2017', *Bulletin of Atomic Scientists*, vol. 73, Issue 5, accessed on 4 July 2017 <http://www.tandfonline.com/doi/full/10.1080/00963402.2017.1363995>

39. Bharat Karnad, *Why India is not a great power (yet)* (New Delhi: Oxford University Press, 2015), p. 387.

40. Vipin Narang, 'Five Myths about India's Nuclear Posture', *The Washington Quarterly*, vol. 36, no. 3 (Summer 2013): 143–57, p. 144 <https://csis-prod.s3.amazonaws.com/s3fs-public/legacy_files/files/publication/TWQ_13Summer_Narang.pdf>

41. Karnad, op. cit., p. 369.

42. Ali Ahmed, 'Indian strategic culture: The Pakistan dimension', in Kanti Bajpai, Saira Basit and V. Krishnappa eds., *India's Grand Strategy: History, Theory, Cases* (New Delhi: Routledge, 2014), p. 298.

43. P. R. Chari, 'India's nuclear doctrine: Confused ambitions,' *The Nonproliferation Review*, Fall–Winter 2000: 123–35, p. 123. Accessed on 1 September 2017 <https://www.nonproliferation.org/wp-content/uploads/npr/73chari.pdf>

44. Sethi, 2009, op. cit., p. 142.

45. Rajesh Rajagopalan, *Second Strike: Arguments about Nuclear War in South Asia* (New Delhi: Viking-Penguin Books India, 2005), p. 22.

46. Vipin Narang, 'Posturing for Peace? Pakistan's Nuclear Postures and South Asian Stability', *International Security*, vol. 34, no. 3 (Winter 2009/10), pp. 38–78 <http://belfercenter.org/publication/19882>

47. Lavoy 2015, op. cit., p. 8.

48. Sethi, 2009, op. cit., p. 251.

49. Sunil Dasgupta, 'The Indian Army: Challenges in the Age of Nuclear Weapons and Terrorism', in Harsh V. Pant (ed.), *Handbook of Indian Defence Policy: Themes, Structures and Doctrine* (London: Routledge, 2015), pp. 129–44, p. 136.

50. Karnad, 2015, op. cit., pp. 366–74.

51. Manoj Joshi, 'Govt Should Not Delay Reforms in the Armed Forces', *Mail Today*, 29 January 2009.

52. Vijay Shankar, 'India–Pakistan–China: Nuclear Policy and Deterrence Stability', *The Strategist*, Institute for Peace and Conflict Studies, 10 March 2014, accessed on 1 September 2017 <http://www.ipcs.org/article/india/india-pakistan-china-nuclear-policy-and-deterrence-stability-4331.html>; Balraj Nagal, 'Strategic Stability—Conundrum, Challenge and Dilemma: The Case of India, China and Pakistan', *Center for Land Warfare Studies Journal* (Summer 2015): 1–22, p. 11 <http://www.claws.in/images/journals_doc/1190813178_BalrajNagal.pdf>

53. Inter-Services Public Relations, 'Press release', 19 April 2011, accessed 15 July 2017 <https://www.ispr.gov.

pk/front/main.asp?id=1721&o=t-press_release>

54. Rajesh Rajagopalan, 'Expanding India's nuclear options,' 10 January 2017, accessed on 5 July 2017 <http://www.orfonline.org/expert-speaks/expanding-indias-nuclear-options/>

55. Manoj Joshi, 'Policy Report: The Unending Quest to Reform India's National Security System', S. Rajaratnam School of International Studies, March 2014, p. 2, accessed on 27 July 2017 <https://www.rsis.edu.sg/wp-content/uploads/2014/09/PR140301_The_Unending_Quest_to_Reform_India_National_Security_System.pdf>

56. Joshi, 2014, op. cit., p. 7.

57. Manmohan Singh, 'Inaugural Address by Dr Manmohan Singh, Prime Minister of India on a Nuclear Weapon-Free World: From Conception to Reality, *IDSA*, 2 April 2014 <http://www.idsa.in/keyspeeches/InauguralAddressShriManmohanSingh>

58. P. R. Chari, 2014, op. cit.

59. Douglas Busvine, 'Modi says committed to no first use of nuclear weapons', Reuters, 16 April 2014, accessed on 8 August 2017 <http://in.reuters.com/article/uk-india-election-nuclear/modi-says-committed-to-no-first-use-of-nuclear-weapons-idINKBN0D20QB20140416>

60. Happymon Jacob, 'The Role of Nuclear Weapons' (pp. 315–38), in Happymon Jacob (ed.), *Does India Think Strategically: Institutions, Strategic Culture and Security Policies* (New Delhi: Manohar, 2014), p. 332.

61. Dinshaw Mistry, 'Military Technology, National Power and Regional Security', in Lowell Dittmar (ed.), *In South Asia's Nuclear Security Dilemma: India, Pakistan and China*, pp. 49–72 (New York: M.E. Sharpe), pp. 70–72.

62. Narang, 2013, op. cit., p. 152.

63. K. Santhanam and Ashok Parthasarathy, 'Pokhran-II: an H-bomb disaster', *Business Standard*, 21 January 2013, accessed on 31 July 2017 <http://www.business-standard.com/article/opinion/k-santhanam-ashok-parthasarathi-pokhran-ii-an-h-bomb-disaster-109121100016_1.html>

64. Vipin Narang, *Nuclear Strategy in the Modern Era: Regional Powers and International Conflict* (Princeton: Princeton University Press, 2014), p. 116.

65. Varghese Koithara, *Managing India's Nuclear Forces* (Washington DC: Brookings Institution Press, 2012), p. 9.

66. Anit Mukherjee, 'Correspondence: Secrecy, Civil-Military Relations, and India's Nuclear Weapons Program', *International Security*, vol. 39, no. 3 (Winter 2014/15): 202–14, pp. 202–3.

67. Prakash, 2014, op. cit., p. 1.

68. Karnad, 2015, op. cit., p. 370.

69. Firdaus Ahmed, 'Nuclear Doctrine: One Gaffe too Many', *India Together*,

14 March 2014, accessed on 17 July 2017 <http://indiatogether.org/gaffes-in-india-s-nuclear-doctrine-op-ed>

70. Rajagopalan, 2005, op. cit., p. 57.

71. P. R. Chari, 2000, op. cit., p. 132.

72. Vipin Narang, 'Did India Change its Nuclear Doctrine?: Much Ado about Nothing', *IDSA*, 1 March 2011, accessed on 30 July 2017 <http://www.idsa.in/idsacomments/DidIndiaChangeitsNuclearDoctrine_vnarang_010311>

73. Bharat Karnad, 'Joint Forces Doctrine—Passive, Defensive Inward-Turned, and Disappointing,' Blog, 30 April 2017, accessed 15 July 2017 <https://bharatkarnad.com/2017/04/30/joint-forces-doctrine-passive-defensive-inward-turned-and-disappointing/>

74. Joshi, 2014, op. cit., p. 8.

75. S. Padmanabhan, *A General Speaks* (New Delhi: Manas Publications, 2005).

76. Ali Ahmed, 'A Disjointed Doctrine: Reviewing the Military's Joint Doctrine', *Economic and Political Weekly*, vol. 52, no. 21 (27 May 2017): 10–12, p. 11.

77. Vijay Shankar, 'A Riddle in a Snare, Does India think strategically? Civil-Military Relations, the Keystone of Strategic Thought,' in Happymon Jacob (ed.), *Does India Think Strategically: Institutions, Strategic Culture and Security Policies* (New Delhi: Manohar Publishers & Distributors, 2014), 217–44, p. 236.

78. Sundarji, 2003, op. cit.

79. D. Scott Sagan, 'Why do states build nuclear weapons?', *International Security*, vol. 21, no. 3 (Winter 1996/97): 54–86, p. 55.

80. Bharat Karnad, *Nuclear Weapons and Indian Security: The Realist Foundations of Strategy* (New Delhi: MacMillan, 2002), pp. 281–445.

81. Aabha Dixit, 'Status quo: Maintaining nuclear ambiguity', in David Cortright and Amitabh Mattoo (eds.), *India and the Bomb: Public opinion and nuclear options* (New Delhi: Bahri Sons, 1996), 53–68, pp. 61–64.

82. K. Subrahmanyam, 'Role of National Power', in K. Subrahmanyam (ed.), *India and the Nuclear Challenge* (New Delhi: Lancers, 1986), 241–71, pp. 256–7.

83. K. Subrahmanyam, 'An Equal Opportunity NPT', *The Bulletin of Atomic Scientists*, vol. 49, no. 5 (June 1993): 37–39, p. 39.

84. Karnad, 2015, op. cit., pp. 379, 385.

85. Abdul Ghafoor Noorani, 'The Doval Doctrine', *Frontline*, 13 November 2015 <http://www.frontline.in/the-nation/the-doval-doctrine/article7800194.ece>

86. Rajat Pandit, 'India cannot rule out possibility of two-front war with China and Pakistan, Army chief General Bipin Rawat says', *The Times of India*, 6 September 2017 <http://timesofindia.indiatimes.com/india/india-cannot-rule-out-possibility-of-two-front-war-with-china-and-pakistan-army-chief-general-bipin-rawat-says/articleshow/60395986.cms>

87. Shashank Joshi, 'India's Nuclear Anxieties: The Debate Over Doctrine', *Arms Control Association*, May 2015, accessed on 5 September 2017 <https://www.armscontrol.org/ACT/2015_05/Features/India-Nuclear-Anxieties-The-Debate-Over-Doctrine>

88. Shashank Joshi, 'Here's why Manohar Parrikar should pause before declaring his 'personal view' on nuclear policy', scroll.in, 11 November 2016. Accessed on 9 September 2017 <https://scroll.in/article/821316/heres-why-manohar-parrikar-should-pause-before-declaring-his-personal-view-on-nuclear-policy>

3

Nuclear Weapons Governance in India: How Robust and Stringent?

Sitakanta Mishra

Nuclearised South Asia is a reality today with reasonable nuclear stability, invalidating the conventional wisdom that the region is on the brink of a 'nuclear winter'. Two decades ago, the proliferation pessimists hyped the Indian subcontinent and the line of control in Kashmir as 'the most dangerous place in the world'.[1] Contrary to their belief, this region has experienced relative nuclear peace and the chances of nuclear war between India and Pakistan seem remote now. Gradually, both countries have worked out many conventional and nuclear confidence building measures (CBMs) to reduce the chances of distrust and misinterpretations in times of crises. This is an offshoot partly of the 'prudent and careful military doctrines' adopted by India and the consequent functioning of 'existential deterrence' between India and Pakistan.[2] The nuclear postures of both countries are based on the 'second-strike' capability—a 'better and more sensible choice than the superpowers did'.[3] India's nuclear doctrine aims at 'minimum nuclear deterrence' with an unambiguous no first use pledge, unlike Pakistan that has adhered to a policy of first-use but as 'last resort ... if Pakistan is threatened with extinction'[4] (a strategy closer to Israel), which is argued to be 'not trigger-happy' as generally portrayed.

However, lesser known is how robust the nuclear command and control mechanism in place, and how stringent the safety-security arrangement to corroborate its aim to maintain a stable South Asian nuclear order are. By drawing data and information from open sources, this chapter attempts to bring to fore the evolution and the debate involving the effectiveness of 'nuclear weapons governance' in India, specifically by analysing the command and control architecture, and safety-security system in place to manage India's nuclear arsenal.

The Magnitude

Any attempt to fathom the exact contours of India's nuclear weapons inventory and related assets would be futile as New Delhi adheres to a great deal of secrecy as a security strategy, unlike the US that publishes in detail the technical safeguards and procedural steps it takes to secure its nuclear assets.[5] Secrecy has been a constant factor ever since India intended to develop 'operational nuclear forces under the hostile gaze of the non-proliferation regime's lead enforcer, the United States'.[6] Further, India's self-imposed NFU posture demands 'opacity and ambiguity' of the steps it takes to ensure nuclear deterrence which are viewed as stabilising factors for nuclear South Asia while opening space for considerable speculation. As India's deterrent (triad) is still in the making, any transparency initiative will have to be limited and certainly on India's terms and confidence in reliability and survivability of its own capability. When its status upgrades from *de facto* to *de jure* nuclear weapon state, or it achieves greater international acceptance, India may afford much more transparency.

For that matter, 'the Indian government has only shown the barest of glimpses of what steps it believes are necessary to ensure deterrence while maximising safety' and security.[7] Except isolated, anecdotal reports, there is no official or authoritative account of India's nuclear weapons inventory and the system of safe-keep in place. *The Bulletin of the Atomic Scientists*, Nuclear Notebook 2017, says 'India is estimated to have produced enough plutonium for 150–200 nuclear warheads but has likely produced only 120–130.'[8] According to the International Panel on Fissile Materials (IPFM),

India's nuclear arsenal is estimated to include 110–120 warheads as of the end of 2014.[9] Its 'stockpile of fissile materials is estimated to include 3.2 ± 0.9 tonnes of HEU, 0.59 ±0.18 tonnes of weapon-grade plutonium, and 5.5±0.4 tonnes of reactor-grade plutonium, that includes 5.1±0.4 of material considered strategic reserve and 0.4 tonnes of safeguarded plutonium. ... The total amount of weapon-grade plutonium stockpile is estimated to be 0.59 ± 0.2 tonnes.'[10] Another study mentions 'the predicted number of weapons made from its weapon-grade plutonium at the end of 2014 is about 97 with a range of 77–123. These values are rounded to 100 nuclear weapons with a range of 75–125 nuclear weapons.'[11]

Concerning the quantity of warheads, without doubt, India's constitute only a fraction of the global inventory. This, however, does not bestow any less responsibility. India has significant reasons to ensure stringent safekeeping of its nuclear infrastructure. The complicated regional security environment, existence of international terror and smuggling networks, and above all, the unique nature of its nuclear program necessitates nuclear governance in India to be a priority. Given India's enthusiastic participation and official statements during all nuclear security summits (NSS), acknowledging the importance of nuclear security, one could say that India is conscious of the fact that 'credible threats exist and nuclear security is important'. 'India fully shares the continuing global concern on possible breaches of nuclear security'[12] and the various global initiatives on nuclear security are also in India's own interest.

The sections that follow, attempt to gather bits and pieces of information from various yards, especially brought to notice by various scholars and retired government officials dealing with nuclear governance issues in India. With a broader view of nuclear weapons control beyond the traditional focus on command and control, one can easily identify India's 'assertive' control of strategic assets within its democratic political framework. With the 'general principles of democratic accountability and civilian control of the security sector to the specific area of nuclear weapons',[13] India's political system acts as the foremost deterrent against misuse or misappropriation of its nuclear assets. Besides, the evolving command and control structure and active and passive defence measures, along with the intelligence system, have turned India's nuclear weapons domain into a fortress.

Command and Control

In fact, India's Draft Nuclear Doctrine 1999 prescribes a force posture to be 'a triad of aircraft, mobile land-based missiles and sea-based assets. ... Survivability of the forces will be enhanced by a combination of multiple redundant systems, mobility, dispersion and deception.'[14] As 'the doctrine envisages assured capability to shift from peacetime deployment to fully employable forces in the shortest possible time, and the ability to retaliate effectively even in a case of significant degradation by hostile strikes', robust command and control and stringent safety-security arrangement is an utter necessity. Admiral L. Ramdas (retd), a former Chief of Staff of the Indian Navy says:

> As of now, there is simply no robust command, control, communications and intelligence system (C3I) in place in either state [India and Pakistan]. Given the economic and technological constraints, this is not likely to materialise for some time to come.[15]

He highlights largely a few non-nuclear 'incidents' and crisis management loopholes to conclude that India's track-record is poor, and refuses to accept the reassurances regarding 'safety' and 'command and control' which have been provided from time to time by government agencies, their nuclear scientists, and also independent analysts.[16] He finds 'the command and control systems...[and] technologies to ensure safety and to prevent accidental deployment of nuclear weapons are still at an embryonic stage'.

According to R. Prasannan, a command and control room had been built for the top 'political leadership' on Prime Minister I. K. Gujral's

(April 1997 to March 1998) instructions.[17] Raj Chengappa in his book *Weapons of Peace* recollects that former Prime Minister P. V. Narasimha Rao (1991–1996) had asked Dr A. P. J. Abdul Kalam, then scientific advisor to the Minister of Defence and the Chairman of the Defence Research and Development Organisation (DRDO), to establish a command and control system on the following four specific principles: (1) the nuclear core should be stored at several strategic sites across the country and not just at Bhabha Atomic Research Centre (BARC), Trombay; (2) arrangements should be made for mating the core with its assembly in the shortest possible time when the need arises; (3) it should be ensured that the command to trigger the bomb remains strictly under civilian control; and (4) the overall system should be so designed that at least three agencies have to combine their efforts if the bomb has to be prepared for a launch.[18] Nothing is known about the steps taken towards the final shape of this directive.

Almost two decades have passed since India proclaimed itself a nuclear weapon state, with a declaratory NFU posture in 1998 and broad outlines of its command and control system in 2003. Even though India's nuclear weapon status has now achieved greater international acceptance, no authoritative detail about the steps it takes to prioritise nuclear weapons safety and security is publicly available. Nothing is known about the physical security arrangements for the strategic nuclear plants except the 2014 Ministry of External Affairs (MEA) document stating that 'Separate institutions and operating procedures exist for nuclear security at India's strategic facilities'.[19]

A certain level of ambiguity in South Asian nuclear discourse is viewed as a stabilising factor but 'unconstrained ambiguity will jeopardise the verifiability and effectiveness of future nuclear-related agreements India and Pakistan may negotiate, as well as those they have already acceded to.'[20] Limited transparency is also necessary for management of crises and long-term regional stability. This does not mean that there is nothing in place concerning command and control, and safety-security of India's nuclear inventory.

Designated Authority

India is known to have developed its nuclear command and control systems, strategic force management mechanisms, and deployment and operational structure over the past two decades. In fact, the design of the warheads and missiles, and the wide range of associated equipment and systems, must be capable of safe and reliable operation for effective performance of the C&C. Even though the President of India, as head of the state, is the Commander-in-Chief of the Indian Armed Forces, the Indian Prime Minister exercises ultimate control over all nuclear weapons. The Prime Minister and the Cabinet Committee on Security (CCS) is designated as the Nuclear Command Authority (NCA) in charge of India's nuclear deterrent. In effect, the command of India's nuclear forces flows from the PMO through the office of the National Security Advisor to the Chairman of the Chiefs of Staff Committee and the C-in-C of the tri-service Strategic Forces Command. More practically, the SFC commander interacts directly with the NSA, the Strategic Planning Staff (SPS), and the NCA, more or less circumventing the conventional military entirely.[21] While the SFC is assigned to execute nuclear operations, the SPS, constituting representatives from the three defence services, science and technology establishments, and other experts from related domains, including the MEA, is mandated to undertake long-range planning and provide independent advice to the NCA.

The NCA is advised by the National Security Council and supported by the SPS. The Strategic Armament Safety Authority (SASA) that functions directly under the NCA is responsible for all matters relating to the safety and security of India's nuclear and delivery assets at all locations. It is assigned the responsibility to review and update storage and transfer procedures for nuclear armaments,

Structure of India's Command and Control

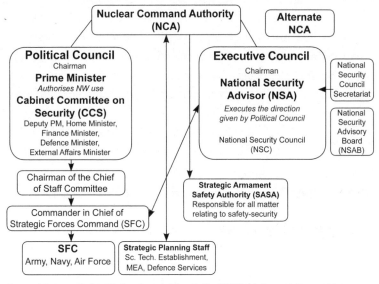

Source: Manpreet Sethi, *Nuclear Strategy*, New Delhi, KW Publishers, 2010, p. 166.

including the submarine-based component. In terms of preparedness, regular drills are conducted to assess the efficiency of the systems to respond to possible escalatory as well as surprise attack scenarios. Reportedly, specialised units have also been deployed for operation in a nuclearised environment.[22]

Mated and De-mated

As per the general perception, India's nuclear warheads are stored in a partially-disassembled, de-mated state (not on hair-trigger alert), with the fissile core kept separate from the physics package and the delivery systems that makes them immune to unauthorised launch during peacetime. However, for India's second-strike posture, 'the most important thing is how fast we can react', maybe within minutes from anywhere at any time.[23] A former head of DRDO was quoted as saying in July 2013 that the DRDO was mandated to work on cannisterised systems that can launch from anywhere at any time for all Indian nuclear missile systems eventually.[24] Even if the entire inventory is not cannisterised yet, one would believe that a part of India's arsenal was kept mated in peacetime. According to Vipin Narang, the belief that India keeps its nuclear weapons in a disassembled state 'is largely now a myth. The "cannisterized" state is a far cry from the prevailing perception that India maintains its nuclear force in a relatively recessed state.'[25] Cannisterised missiles are mated and if carried or loaded, they require special safety-security arrangements. No authoritative information is available about the level of safety-security sophistication India practices regarding its mated arsenal.

International security observers and experts believe that increasing numbers of strategic nuclear forces in South Asia make safety, security, and control issues far more problematic. Specifically, the requirement to keep warheads and delivery systems separate for reasons of security and control could add to design and maintenance problems.[26] According to Verghese Koithara, if India's missiles are to be deployed in a genuinely mobile manner, then the multi-channel system of nuclear weapons control can be very cumbersome. More importantly, as India's current missile force is primarily land mobile and thus requires decentralised operations, it can be prone to logistic and security challenges.[27] Though, according to experts from DRDO, 'adequate safety provisions are made through electromechanical devices' like safety arming and detonation mechanisms of guided missiles 'to prevent accidental initiation'.[28]

Ensuring Custody

It would be safe to assume that India has already taken, and is continuously striving for, extra measures to achieve maximum level of safety and security of its nuclear warheads. New Delhi is believed to have developed its own Permissive Action Link (PAL) mechanisms and the modern safety designs (enhanced nuclear detonation safety system—ENDS) to block critical arming signals specifically designed to prevent unauthorised use and protect electrical systems respectively. According to Raj Chengappa, 'a series of half a dozen safety locks' ensure that Indian nuclear warheads can explode only when desired.[29] 'It could be assumed that these safety locks are based on PAL technologies.'[30] In a warlike situation, the assembly of dispersed components would come under extreme stress, which would demand very high managerial and material safety and security competencies.

As far as the custody of nuclear warheads is concerned, the authority is not centralised at the SFC; rather, it is 'divided between two civilian scientific agencies: the BARC and the DRDO, which control the fissile cores and non-nuclear firing assemblies'. Maintenance of the weapons is also their responsibility. In the same manner, maintenance of delivery vehicles would be undertaken by respective services. The C-in-C of SFC exercises control over the strategic units which actually store, maintain, and deploy the delivery mechanisms.[31] 'The SFC provisions the primary and alternative command posts, operations rooms and communication links and maintains an interface with the Atomic Energy Commission (AEC) and DRDO.'[32] However, 'the air force and the navy each retain separate control of nuclear delivery systems.'[33]

Military in the Loop

From its inception, 'the armed forces remained out of the nuclear decision making loop' in India.[34] Till 1998, there were allegations by high-ranking defence officials that they had neither been told about India's nuclear weapons programme nor consulted about the size and shape of the nuclear deterrent or the contours of India's nuclear doctrine. Only the engineer regiment of the army was known to have been involved in digging shafts for nuclear tests and in providing certain logistics support. But sometime in 1986, the then Chief of Air Staff, 'Air Chief Marshal S. K. Mehra was inducted to work on the weapons delivery aspects of the nuclear programme.'[35] One can deduce that the role of the armed forces in India's nuclear weapons programme was rather sporadic or intermittent, and was on 'need to know' basis. The Arun Singh Committee subsequently recommended inclusion of Service chiefs into the nuclear decision making loop and distributing the task of warhead assembly to many different agencies to minimise the dangers of an inadvertent or accidental detonation.[36] Basically, to project a credible deterrent, what India's nuclear doctrine prescribes for, the armed forces have to be involved. Post-1998, according to Ambassador Arundhati Ghose, the military seems 'much closer into the nuclear decision making process in India.'[37]

Still Assertive?

After the Indo-US civil nuclear deal, separation of civil nuclear facilities from facilities that are associated with its strategic programme must have smoothened streamlining the safety-security arrangements. But, with the introduction of the third leg of India's nuclear triad, the problem of unauthorised launch becomes a complex technological question. In regard to deployed submarines, two related command and control issues are: (1) the 'always-never dilemma' (nuclear weapons always ready for use but can never be launched accidentally or unauthorised),[38] and (2) the problem of maintaining communications. 'India has not yet explained how it intends to retain active civilian control over its Submarine-Launched Ballistic Missile (SLBM) arsenal.'[39] For a sea-based asset away from shores with launching control delegated to seniority on board the vessel, the existing command and control model is not fully applicable. As a designated 'second-strike' capacity asset, the NCA is viewed to be unable to effectively and credibly implement fail-safe measures on assets on board the vessel. With the commissioning of Arihant,[40] the nuclear submarine SSBN with ready nuclear systems available with the commander, has the command and control structure originally based on assertive civilian control been silently revisited? If so, 'whose finger is on the nuclear trigger at sea?'[41]

No movement is observed since the BJP-led National Democratic Alliance coalition pledged in its 2014 election manifesto to 'study in detail India's nuclear doctrine, and revise and update it, to make it relevant to challenges of current times'.[42] Yet, it would not be realistic to presume that New Delhi has not yet adjusted or devised the required control structure for its third leg of the triad within either the existing doctrinaire position and command and control system, or tweaking it suitably. Taking cue from other nuclear weapon states with SLBMs, India might have resolved this issue either through technological means like the USA did, replacing its 'two-man rule', or following the British principle of 'beyond the grave' pre-planned instructions to their submarine commanders.[43] If not, it would be interesting to know if India has invented any innovative system to govern the third leg of its nuclear triad within the original system in place.

In any case, in the near-term, India would adopt the 'bastion strategy', i.e. operating the SSBN close to its territorial waters within the protective envelop of its land- and sea-based firepower. Given the nascent status of the program, the submarines may not perform truly autonomous operations immediately. As India's nuclear deterrence is conditioned by the 'China factor', gradually, the third leg of its nuclear triad has to attain more credibility with trans-oceanic capabilities. As per recent reports, India is working on a classified project to build 'six nuclear-powered attack submarines witnessing China's naval build-up and its increasing military manoeuvring in the Indo-Pacific region.'[44] The Indian Navy is gearing up to play a bigger role, including under the proposed quadrilateral coalition between India, USA, Australia, and Japan.[45] In the long-term, therefore, the critical issue of command and control of SSBNs might surface which can be addressed with sophisticated technological approach as it evolves.

Alternate NCA

One less debated aspect of India's Command and Control discourse is the Alternate NCA (ANCA) or alternate chain of command in place. The CCS on 4 January 2003 reviewed the operationalisation of India's nuclear doctrine and 'approved the arrangements for alternate chains of command for retaliatory nuclear strikes in all eventualities.'[46] In case the Political Council is obliterated or immobilised due to first strike by adversary, there are alternate, pre-decided councils that will replace the Political Council and fulfil the NCA's mission.[47] It is expected that 'chains of constitutional and nuclear succession that best fit India's political institutions and strategic situation' are in place.[48] But not an iota of information is available on the contours of India's ANCA, obviously for security reasons. One can argue, 'While announcing details about the NCA, the CCS prudently attempted to strike a balance between transparency—assuring the world of civilian primacy and public accountability; and secrecy—protecting alternative chains of command and, thereby, strengthening nuclear deterrence.'[49] Effectiveness of Indian deterrent rests undoubtedly on the capability to survive the first strike and retaliate with massive force to cause 'damage unacceptable to the aggressor'. Therefore, robustness and stringency are inbuilt into the nuclear strategy since its inception, firstly in the form of secrecy, and gradually striking a balance between declaratory doctrinal position and secretive chain of command.

Sketchy Passive Defence

Passive defensive capabilities in the form of bunkers, protective shelters, etc. have reportedly been built. According to a report by Chandan Nandy in *Hindustan Times* (21 September 2003), the NCA took a decision in 2003 to build bunkers for top decision makers within the South Bloc/PMO, as well as a second bunker to be 'set up within a 400-kilometre (250-mile) radius of the capital…. Potential locations in the northern states of Uttar Pradesh and Rajasthan and in the central province of Madhya Pradesh were being studied by the head of the air force, Air Chief Marshal S. Krishnaswamy, and the chief of the nuclear forces, Air Marshal Teja Mohan Asthana', the first C-in-C of SFC.[50] 'Details about these arrangements are understandably sketchy and insufficient to make judgements about the adequacy of these efforts or their weaknesses.'[51]

All Spectrum Missile Defence

To supplement the passive defence measures to enhance the safety, security, and survivability of NCA, India is pursuing a ballistic missile defence (BMD) programme. Reportedly, India is hurrying up the deployment of a BMD system to stave off threats from ballistic missiles, which was supposed to be ready for deployment by the end of 2016.[52] Earlier in 2012, media reports revealed that Delhi and Mumbai were chosen for BMD systems 'that can be put in place at short notice. To ensure maximum protection against air-borne threats, DRDO will put up a mix of counter-attack missiles which will be able to shoot down enemy missiles both within earth's atmosphere (endo-atmospheric) and outside it (exo-atmospheric).'[53] A recent report says that India has decided to install BMD systems at two villages in Alwar and Pali in the western state of Rajasthan, located less than 800 kilometres from the Pakistani capital Islamabad.[54]

The threat of ballistic missiles and nuclear weapons does indeed remain a great concern in South Asia, but is not the only form of contingency that India will experience. One perceived strategic implication of the (nuclear) cruise missiles, especially with Pakistan, is that it has 'lowered the index of stability in the region'.[55] A combination of ballistic and cruise missile threat could severely complicate and stress India's deterrent. Though India has achieved substantial headway in BMD, it cannot hope for a credible defence unless its missile defence strategy addresses both ballistic and cruise missile threats. Theoretically, defence against cruise missiles is possible.

Using anti-air missiles of various ranges, it may be possible to intercept anti-ship cruise missiles (although intercepting land-attack missiles remains a Herculean task). Should India not plan for a cruise missile defence (CMD) programme in the long run? This is, however, not an easily achieved capability. The US Department of Defence (in the study named 'Integrated Air and Missile Defence') has stated that it is not yet fully successful and has identified at least nine capability gaps in the US CMD programme. Most prominent among them is the inability to detect small, low-speed, low-altitude targets.[56] When advanced nations have not been able to develop such a system yet, can India afford to, and can it overcome the technical hurdles in achieving such a technology-intensive system to counter the emerging cruise missile threat effectively?

Safety-Security Intricate

Safety-security of nuclear and delivery vectors involves intricate inter-relationship between personnel, information management, organisational coordination, safety, and security components. So far, there is no reported significant incident/ event leading to a nuclear disaster in India. Suffice it to say, that India has maintained a stringent personnel reliability framework in all its nuclear-related institutions, prescribing a code of ethics and conduct to be followed. Besides the defence establishments' own mechanisms and procedures of scrutiny of their own staff and the SFC's standard operating procedures, the intelligence agencies and police network are part of the mechanism to ensure the integrity of the personnel employed in this sector.

Regarding transportation of nuclear assets, specific security levels are specified for different materials (Category 1 to 5), depending upon their degree of fissile characteristics and danger involved. This includes, among others, prior approval for the shipment, special vehicle, security locks, appropriate training of personnel involved, additional security and escort by armed guards, secure communication support, and an online tracking system, etc.[57] But in a confrontational situation, a decision to move and deploy adds to the complexity of challenges for which India is expected to be prepared.

In addition, the intelligence system to monitor, gather information, and analyse comprises both civilian and military agencies such as the National Technical Research Organisation (NTRO), Defence Intelligence Agency (DIA), which oversees signal intelligence, Defence Imagery Processing and Assessment Centre, Research and Analyses Wing (R&AW), Technical Coordination Group (TCG), and the Intelligence Coordination Group (ICG), with their impeccable records.[58]

An Introspection

'Command and control of nuclear arsenals is a complex subject and involves many issues',[59] actors, and factors. India's command and control, and related safety-security arrangement, has evolved within a challenging nuclearised environment. The command and control system may not anymore be the exclusive domain of the civilian authorities today when the role of military is an imperative to convey the adversary unambiguously the effectiveness of its deterrent. Therefore, it is a matter of introspection whether India's nuclear command and control is still highly politically 'assertive'.

In a vibrant democratic set up like India, exercising democratic governance on security apparatus has always been and will remain a uniquely fraught issue. There will always be an inherent tension between the needs for transparency, accountability, oversight in a democracy, and the requirements for secrecy and swift, executive decisions in the realm of national security. As controlling the nuclear bomb required extraordinary and unprecedented security measures (especially secrecy with the 'born secret' doctrine), governing the atom will remain inherently a challenge to the norms and values of a democratic society.[60] Imbibed with a liberal democracy that inherits an 'argumentative' culture, India's case is no exception.

However, if one observes the totality of India's nuclear weapons governance, the central principle and core element is that the 'doctrine' it has adopted decides the nuclear 'strategy'; the nuclear strategy in turn decides the nature and substance of the command and control system which is still evolving to execute the strategy in accordance with that doctrine.[61] There is a strong linkage between India's nuclear doctrine, strategy, and structure with ample scope for robustness and stringency.

Notes and References

1. Jonathan Marcus, 'Analysis: The World's Most Dangerous Place?', *BBC*, 23 March 2000, accessed at <http://news.bbc.co.uk/2/hi/south_asia/68/021.stm>

2. Rajesh Rajagopalan, *Second Strike: Arguments about Nuclear War in South Asia* (New Delhi: Penguin/ Viking, 2005).

3. Ibid.

4. Former Pakistani President General Pervez Musharraf quoted as having said this in an interview published in April 2002 in the German magazine, *Der Spiegel*; Peter R. Lavoy, 'Pakistan's Nuclear Posture: Security and Survivability', accessed at <http://www.npolicy.org/article_file/Pakistans_Nuclear_Posture-Security_and_Survivability.pdf>

5. According to a recent report, 'The Pentagon has thrown a cloak of secrecy over assessments of the safety and security of its nuclear weapons operation.... The stated reason for the change is to prevent adversaries from learning too much about US nuclear weapons vulnerabilities. Navy Capt. Greg Hicks, spokesman for the Joint Chiefs of Staff, said the added layer of secrecy was deemed necessary'. Robert Burns, 'US Tightens Security, Secrecy Over Nuclear Weapons Operations', accessed at <https://www.thestar.com/news/world/2017/07/04/us-tightens-security-secrecy-over-nuclear-weapons-operations.html>

6. Gaurav Kampani, 'New Delhi's Long Nuclear Journey', *International Security*, vol. 38, no. 4 (Spring 2014), p. 81.

7. Christopher Clary, 'Lift the Lid off Nuclear Secrecy', *BusinessLine*, *The Hindu*, 15 July 2013, accessed at <http://www.thehindubusinessline.com/opinion/lift-the-lid-off-nuclear-secrecy/article4917883.ece>

8. Hans M. Kristensen and Robert S. Norris, 'Indian nuclear forces, 2017', *Bulletin of the Atomic Scientists*, vol. 73, no. 4 (2017), pp. 205–9, accessed at <https://doi.org/10.1080/00963402.2017.1337998>

9. International Panel on Fissile Materials, 5 August 2016, accessed

at <http://fissilematerials.org/countries/india.html>

10. Ibid.

11. David Albright and Serena Kelleher-Vergantini, 'India's Stocks of Civil and Military Plutonium and Highly Enriched Uranium, End 2014', 2 November 2015, p. 17, accessed at <https://isis-online.org/uploads/isis-reports/documents/India_Fissile_Material_Stock_November2_2015-Final.pdf>

12. 'Khurshid: Nuclear Terrorism Serious Threat to Global Peace', *Indian Express*, 26 March 2014, accessed at <http://indianexpress.com/article/india/india-others/khurshid-nuclear-terrorism-serious-threat-to-global-peace/>

13. Hans Born, Bates Gill, and Heiner Hanggi (eds.), *Governing the Bomb: Civilian Control and Democratic Accountability of Nuclear Weapons* (SIPRI/Oxford University Press, 2010), p. viii.

14. 'Draft Report of National Security Advisory Board on Indian Nuclear Doctrine', 17 August 1999, accessed at <http://mea.gov.in/in-focus-article.htm?18916/Draft+Report+of+National+Security+Advisory+Board+on+Indian+Nuclear+Doctrine>

15. L. Ramdas, 'Myths and Realities of Nuclear Command and Control in India and Pakistan', *Disarmament Diplomacy*, Issue No. 54, February 2001.

16. Ibid.

17. R. Prasannan, 'Secret? Tell it to the Press', *The Week*, 9 August 2015.

18. Raj Chengappa, *Weapons of Peace:*

The Secret Story of India's Quest to be a Nuclear Power (New Delhi: Harper Collins Publishers India, 2000), p. 391.

19. Ministry of External Affairs (MEA), 'Nuclear Security in India', accessed at <http://www.human.ula.ve/catedralibreindia/documentos/india_nuclear.pdf>.

20. Gaurav Rajen and Kent Biringer, 'Nuclear-Related Agreements and Cooperation in South Asia', *Disarmament Diplomacy*, Issue No. 55, March 2001, accessed at <http://www.acronym.org.uk/dd/dd55/55rajen.htm>.

21. Gaurav Kampani, 'India's Evolving Civil-Military Institutions in an Operational Nuclear Context', 30 June 2016, accessed at <http://carnegieendowment.org/2016/06/30/india-s-evolving-civil-military-institutions-in-operational-nuclear-context-pub–63910>.

22. Shyam Saran, 'India's Nuclear Weapons not for National Pride', Research and Information System for Developing Countries, accessed at <http://ris.org.in/images/RIS_images/pdf/tribune–9may%202013.pdf>

23. Shiv Aroor, 'My job is to speed up India's nuclear strike time, DRDO chief says', *Daily Mail*, 3 July 2013 accessed at <http://www.dailymail.co.uk/indiahome/indianews/article-2355107/Dr-Avinash-Chander-My-job-speed-Indias-nuclear-strike-time-DRDO-chief-says.html>

24. Ibid.

25. Vipin Narang, 'Five Myths about India's Nuclear Posture', *The Washington Quarterly*, vol. 36, no. 3, 2013, pp. 143–57.

26. Verghese Koithara, *Making India's Nuclear Forces* (New Delhi: Routledge, 2012), p. 127.

27. Ibid.

28. T. V. Karthikeyan, A. K. Kapoor, *Guided Missiles*, Defence Research and Development Organisation, Ministry of Defence, Government of India, p. 16, accessed at <http://drdo.gov.in/drdo/data/Guided%20Missiles.pdf>

29. Raj Chengappa, *Weapons of Peace: The Secret Story of India's Quest to be a Nuclear Power* (New Delhi: Harper Collins Publishers India, 2000), p. 391.

30. Clary, n. 7

31. Manpreet Sethi, *Nuclear Strategy: India's March Towards Credible Deterrence* (New Delhi: KW Publishers Pvt. Ltd., 2009), p. 169.

32. Ali Ahmed, 'Indian Nuclear Command and Control—II', *Aakrosh*, 13 July 2011.

33. Kampani, n. 20.

34. Gurmeet Kanwal, 'Safety and Security of India's N-Weapons', IDSA, accessed at <https://www.idsa-india.org/an-apr–1.01.htm>

35. Ibid.

36. Raj Chengappa, 'End the Wink and Nudge Approach', *Weapons of Peace: The Secret Story of India's Quest to be a Nuclear Power* (New Delhi: Harper Collins Publishers India Pvt. Ltd., 2000), pp. 365–6.

37. Ayesha Ray, *The Soldier and the State in India: Nuclear Weapons, Counterinsurgency, and the Transformation of Indian Civil-Military Relations* (New Delhi: Sage, 2013), p. 93.

38. P. D. Feaver, 'Command and Control in Emerging Nuclear Nations', *International Security*, vol. 17, no. 3 (1992–93), pp. 168–9.

39. Diana Wueger, 'India's Nuclear-Armed Submarines: Deterrence or Danger?' *The Washington Quarterly*, vol. 39, no. 3 (2016), p. 87, accessed at <https://www.tandfonline.com/doi/abs/10.1080/016366 0X.2016.1232636>

40. Nayanima Basu, 'Nuclear submarine INS Arihant quietly commissioned into service', 18 October 2016, accessed at <http://www.thehindubusinessline.com/economy/ins-arihant-commissioning/article9235728.ece>

41. W. P. S. Sidhu, 'Whose finger on the nuclear trigger at sea?', Livemint.com, accessed at <http://www.livemint.com/Opinion/FesGy5sItj3WTJywQdfiKO/Whose-finger-on-the-nuclear-trigger-at-sea.html>

42. 'BJP Election Manifesto 2014', p. 39, accessed at <http://www.bjp.org/images/pdf_2014/full_manifesto_english_07.04.2014.pdf>

43. Sidhu, n. 41.

44. Namrata Tripathi, 'India working on project to build 6 nuclear-powered attack submarines', MSN.com, accessed at <https://www.msn.com/en-in/news/newsindia/india-working-on-project-to-build–6-

nuclear-powered-attack-submarines/
ar-BBG2ay8?li=AAggbRN&ocid=m
ailsignout>

45. 'Quadrilateral security dialogue:
India, Australia, Japan, US hold
talks on Indo-Pacific cooperation',
The Times of India, 12 November
2017.

46. 'The Cabinet Committee on Security
Reviews Operationalization of
India's Nuclear Doctrine', 4 January
2003, accessed at <http://mea.gov.
in/press-releases.htm?dtl/20131/Th
e+Cabinet+Committee+on+Securit
y+Reviews+perationalization+of+In
dias+Nuclear+Doctrine>

47. Nuclear Command Authority,
globalsecurity.org, accessed at
<https://www.globalsecurity.
org/wmd/world/india/
nuclearcommandauthority.htm>

48. Nitin Pai, 'The lines of nuclear
succession', 11 February
2009, accessed at <http://
www.livemint.com/Opinion/
chrIFJKeoja7YzlLnrdwXM/The-
lines-of-nuclear-succession.html>

49. Ibid.

50. AFP, 'India to build two nuclear-
proof bunkers to shield top
leaders: report', accessed at
<http://www.propagandamatrix.
com/220903indiabunkers.html>

51. Rajesh Rajagopalan, 'India: The
Logic of Assured Retaliation', in
Muthiah Alagappa (ed.), *The Long
Shadow: Nuclear Weapons and
Security in 21st Century Asia* (Palo
Alto: Stanford University Press,
2008), p. 202.

52. Rahul Singh, 'India to deploy

defence against ballistic missiles
by 2016, says DRDO chief', *The
Hindustan Times*, 16 September
2014.

53. PTI, 'Missile defence shield for
Delhi, Mumbai planned', accessed at
<http://www.thehindubusinessline.
com/economy/policy/missile-
defence-shield-for-delhi-mumbai-
planned/article3565834.ece>

54. 'India to Deploy Two Ballistic
Missile Defense Systems Near
Pakistan Border', 7 August 2017,
accessed at <https://sputniknews.
com/asia/201708071056252187-
india-missile-defense-systems/>

55. C. Uday Bhaskar quoted saying at
an Asia-Pacific security seminar on
India's Maritime Security Challenges
at the East-West Centre on 7 May
2013, 'Taking a leaf from China,
Pakistan seems to be investing in
cruise missiles. This has lowered
the index of stability in the region',
accessed at <http://indiatoday.
intoday.in/story/cruise-missiles-
china-pakistan-stealth-capabilities-
uday-bhaskarindia-maritime-
security/1/269879.html>

56. John Liang, 'DoD Finds Cruise
Missile Defense "Gaps"',
accessed at <http://military.com/
features/0,15240,110199,00.html>

57. 'Security of Radioactive Material
During Transport,' AERB Safety
Guide No. AERB/NRF-TS/
SG–10, AERB, January 2008.
Available at <http://www.aerb.
gov.in/T/PUBLICATIONS/
CODESGUIDES/sg–10.pdf>

58. Ali Ahmed, 'Indian Nuclear

Command and Control—II', 13 July 2011, accessed at <http://www.indiandefencereview.com/spotlights/indian-nuclear-command-and-control-ii/>

59. Jasjit Singh, 'Nuclear Command and Control', accessed at <https://www.idsa-india.org/an-may–1.01.htm>

60. 'Nuclear Governance and Legislation in Four Nuclear-Armed Democracies: A Comparative Study', accessed at <https://www.nonproliferation.org/wp-content/uploads/2017/09/nuclear-governance-and-legislation-in-four-nuclear-armed-democracies.pdf>

61. Singh, n. 59.

4

India's Export Control Regime

Zafar Ali

Until 1992, India had a rudimentary regulatory control system over dual-use goods and technologies whereby imports/exports were largely governed by the Import and Export Control Act of 1947, which was the principal legal basis for India's strategic trade control system. India's desire to attain a major world power status intensified as the Cold War ended and dividends from the economic reforms and trade liberalisation policies started to increase. A sustained democratic political system, a thriving industry especially the IT sector, and improved military strength rejuvenated India's urge to play a larger role at the world stage.

Munir Akram, former Pakistan Ambassador to the UN says, 'The election of Narendra Modi as Prime Minister and geopolitical developments, particularly the US pivot to Asia and Russia's new Cold War with the West, have revived India's prospects of achieving great power status.'[1] Recognising India's importance for commercial and geostrategic interests in the region, the US has sought to enhance its partnership with India in multifarious fields. Impetus for this new-found friendship emerged in the early 1990s, following India's economic reforms. President Clinton's India visit in 2000 further cemented the ties. As part of Next Steps in Strategic Partnership, in 2004 both the states 'agreed to expand cooperation in three specific areas: civilian nuclear activities, civilian space programmes, and high-technology trade.'[2]

On its part, India regards the US essential for its economic and military capacity building as a major source of capital and technology. In part, it was China's growing power that heightened India's security concerns and it saw an advantage in tilting towards the US in order to improve its relative position. India sees value in nurturing a strategic partnership with the US to acquire high technology for industrial development and move from being a consumer to a supplier country. Identifying critical areas of India's preference for future cooperation, Brian Shoup and Sumit Ganguly say, 'A critical area of future focus lies in the area of high-technology trade, particularly in those technologies that advance India's interests in energy security, aerospace, and nuclear safety.'[3]

To stimulate greater economic activity, create space for ambitious defence modernisation plans, break free from nuclear isolation, and nurture aspirations of major power status, India started to align its import/export policies with international best practices, especially those of the multilateral export control regimes (i.e. Nuclear Supplier Group, Missile Technology Control Regime, Australia Group, and Wassenaar Arrangement), introduced structural changes, and promulgated laws to streamline the import and export control system. This chapter attempts to chart the history and evolution of India's strategic export control regime, take a critical look at the developments of institutional and legislative frameworks as India treaded the path to achieving economic capacity that accentuated the need for foreign technological assistance, outline the licensing and enforcement system for dual-use items, discuss India's aspirations to gradually come out of nuclear isolation, and in the final part, identify the strengths and weaknesses of India's export control regime as it moves to align itself with international best practices.

Overview of India's Export Control Regime

Many experts profess four fundamental requirements (pillars) of an effective export control system: legislation that provides the legal basis, licensing system that sets out licensing conditions and procedures under which licenses are approved or denied, enforcement mechanism for implementation of the law, and government/industry cooperation for enhanced compliance. National commitment forms the core of the whole structure in order for it to be effective. Promoting awareness and self-regulation through a comprehensive strategy of reaching out to the industry, academia, research institutes, governmental agencies, and exporters etc. is essential for effective administration of export controls. The economic, political, and security priorities set by the government therefore, form the basis for export control policies and practices.

India boasts of always being a responsible nuclear power with an 'impeccable' non-proliferation record. However, legislative, regulatory, and administrative measures to regulate the export of dual-use items usable in weapons of mass destruction (WMDs) in the initial period was scant. Dr Rajiv Nayan, Senior Visiting Research Associate at Kings College London and Senior Associate at the Indian Institute of Defence Studies and Analysis, claims:

> by and large, the Indian export control arrangement, as it existed initially, was rudimentary in nature and structure. Although a sense of the importance of responsible controls existed, the Indian government largely viewed export controls within the context of the technology denial regime.[4]

Since 1992, India's strategic export control system began developing in a relatively more structured way yet the basic objective was to facilitate imports and exports, and limited numbers of dual-use items were subjected to export license. According to Indian officials, the first list of dual-use items, i.e. Special Materials, Equipment, and Technologies (SMET) (later revised and notified as SCOMET in 2001), was notified in 1995 though rules for manufacture, use, import, export, and storage of hazardous micro-organisms, and lists of dual-use chemicals subject to regulatory controls were notified in

1989 and 1993 respectively.[5] Identical views were expressed by another Indian official in a subsequent seminar at Tokyo in February 2016. This illustrates that there was little or no serious attention to regulate or monitor export of sensitive dual-use items prior to 1995 despite recognising the sensitivity associated with exports of certain items. Nevertheless, prohibitions and control over nuclear goods, technology, and services, and restriction on disclosure of certain information were established as early as 1962 through the Atomic Energy Act. The regulatory framework for conventional arms, ammunition, and related manufacturing machinery and accessories had been established under the Arms Act 1959 and other legal instruments, which were further elaborated to show compliance with the Wassenaar Arrangement munitions and dual-use lists, Category-6 of the Special Chemicals, Organisms, Materials, Equipment, and Technologies (SCOMET) list, marked as reserved, was inscribed with conventional munitions items. In 2005, India entered into a civil nuclear cooperation agreement with the US and subsequently adopted the Weapons of Mass Destruction and their Delivery Systems (Prohibition of Unlawful Activities) Act–2005. Simultaneously, it also announced its intention to adhere to the Nuclear Suppliers Group (NSG) Guidelines, though formal adherence, as required vide NSG procedural arrangements, was declared in 2016 at the time of submitting its membership application.

Legislative Framework of Export Control Regime and the Evolution of Control List

To institute an effective export control regime, a country first needs to make a political commitment to non-proliferation. Secondly, it needs to have a comprehensive legal, regulatory, and administrative framework that provides for an appropriate implementing authority with adequate powers, and establishes clear jurisdiction over territory, transactions, people, goods, and related technologies (control lists) including criminalisation of offences and penalties. Indian officials and non-proliferation pundits cite PM Nehru's 1947 statement, given below, as the basis for future controls over strategic exports:

> Export is not merely a financial matter. It has international implications … It is desirable for the Government of India to prohibit the export of monazite and thorium nitrate from India … and this would mean that any export would be in accordance with the explicit permission of the Government of India and subject to the conditions laid down.[6]

Notwithstanding the aforementioned high-level political commitment to strategic export control, actual progress towards a workable strategic export control system remained stagnant. India is a nuclear weapon state but not a party to the nuclear Non-Proliferation Treaty (NPT), and has the ability to produce nuclear and missile items. Despite high-level political commitment enshrined in Prime Minister Nehru's proclamation, much attention had not been given to legislative, administrative, and implementation aspects of export controls over strategic and dual-use goods until the promulgation of the Foreign Trade (Development and Regulation) Act 1992. To seek a significant position in global business and keep pace with globalisation and economic integration, India had to streamline its trade policies, frame new laws, and introduce or amend regulations governing foreign trade. 'The laws and facilitations announced by the government were not only related to export and import of goods and services but were also directed at up-gradation of technology and integration of all the departments by using latest technologies available.'[7] Today, India has a whole host of enactments supported by rules, regulations, policies, and orders to administer export controls. The major benchmarks of its legislative developments since 1947 are chronicled below:

a. Import and Export Act 1947: This law was inherited from the British and is considered to be the first permanent enactment that came into effect in March 1947. It vested wide ranging powers in the central government to prohibit and/or control or restrict imports and exports.

b. Atomic Energy Act 1962: This Act prohibits the export and import of designated substances, equipment, plants designed for the production, development, and use of atomic energy or related research, and radioactive substances, unless the transaction is licensed by the designated government authorities.[8] A list of items and substances that are subject to export license from the Department of Atomic Energy (DAE) has also been notified.

c. Customs Act 1962: The British era Sea Customs Act 1878 was replaced by the Customs Act 1962 providing for controlling imports and exports. It empowers Customs to enforce export controls and deal with violations.

d. Foreign Trade (Development and Regulation) Act 1992 (FTDR Act-1992): To adjust to the new economic realities and the changing global trade environment, the Foreign Trade (Development and Regulation) Ordinance 1992 was promulgated by the President in June 1992. The same year, in August, the Ordinance was converted into the Act. The Act vests import and export regulatory powers in the Central Government and 'provides for the development and regulation of foreign trade by facilitating imports into India and augmenting exports from India.'[9] It enables the Central Government to formulate import and export policy, regulate foreign trade, and prohibit, restrict, or otherwise regulate import and export of goods, and to appoint Director General Foreign Trade (DGFT) for implementation of the FTDR Act. The Act inter alia provides for a Foreign Trade Policy and Procedures on export control, to specify the items subject to export control and licensing. The Foreign Trade (Regulation) Rules 1993 and Foreign Trade Policy were subsequently notified. DGFT is responsible for implementing the Foreign Trade Policy or Exim Policy with the main objective of promoting Indian exports.[10] Significant additions related to transit and trans-shipment controls over dual-use goods were made in the Foreign Trade Act in 2010.

e. The Weapons of Mass Destruction and their Delivery Systems (Prohibition of Unlawful Activities) Act 2005. In part, this Act was promulgated under US pressure and the obligatory requirement of the United Nations Security Council Resolution 1540, and in part, to align itself with the requirements of Multilateral Export Control Regimes (MECRs) to bolster credentials for membership of these regimes and become eligible for access to civil nuclear technology. This Act expands governmental authority to regulate export, re-export, re-transfer, transit, and trans-shipments of goods and technologies that could be used for the development, production, handling, operation, maintenance, storage, or dissemination of WMDs or missile delivery devices.[11] As stated in the Act, it provides integrated legal measures to exercise controls over the export of materials, equipment, and technologies, and prohibits unlawful activities in relation to WMDs and their delivery systems, especially prohibiting the possession, manufacture, transportation, acquisition, and development of nuclear weapons, chemical weapons, or biological weapons by non-State actors. As stated in India's official communique to the International Atomic Energy Agency:

> The Act updates the present system of export controls in India with a view to making it more contemporary, by introducing transit and trans-shipment controls, re-transfer provisions, technology transfer controls, brokering controls, and end-use based controls. The Act prohibits the export of any good or technology from India if the exporter knows that it is intended to be used in a WMD programme.[12]

Many of the precepts which had entered the Western countries' export control systems such as catch-all, deemed export, intangible technology, brokering, transit, and trans-shipment etc. were incorporated in the Act. Corresponding changes were also made in the FTDR and Foreign Trade Policy to bring these in line with the spirit of the WMD Act.

There are numerous other legal and statutory codifications that govern the import and export of other dual-use goods, materials, and technologies, for example the Chemical Weapons Convention Act 2000, Explosives Act 1908, Arms Act 1959, Narcotic Drugs and Psychotropic Substances Act 1985, Atomic Energy (Radiation Protection) Rules 2004, Environmental Protection Act 1986, and The Unlawful Activities (Prevention) Amendment Act 2004. Revised guidelines for nuclear transfers were notified on 28 April 2016 just before India submitted its candidacy for the NSG. Indian authorities claim that these are harmonised with the current NSG Part-I Guidelines of June 2015 and encompass prohibition on exports for development of nuclear explosive devices, physical protection requirements set out in the International Atomic Energy Agency (IAEA) INFCIRC/225, application of IAEA safeguards, the right to apply additional conditions of supply, special controls on sensitive exports including enrichment facilities, equipment, and technology, and controls on re-transfer and non-proliferation principle.

SCOMET

The Small Group on Strategic Export Controls was constituted in 1993 with the mandate to develop a list of goods to be subjected to export licensing. Accordingly, a list of special materials, equipment, and technologies was developed, which was published for the first time in the Export Import Policy Order in 1995. Separately, the Department of Atomic Energy had developed a list under the Atomic Energy Act 1962 containing nuclear-related materials which are subject to export license by DAE.[13] Another Small Group on Strategic Export Controls, constituted by DGTR in 1999, was tasked to review the existing lists and procedures; consequently, the original SMET list was made more comprehensive by the addition of entries on 'chemicals' and 'organisms'—describing the revised list as Special Chemicals, Organisms, Materials, Equipment, and Technologies. SCOMET is notified in the Export Policy in Schedule 2 Appendix 3 of the Indian Tariff Classification (Harmonised System)—ITC (HS)—Classifications of Export and Import Items. 'The revised SCOMET list became effective from 1 April 2000. The export of items in this list is prohibited or permitted only under license.'[14] Category 6 of the SCOMET list, initially kept as reserve, has recently been populated with the Munitions List as India was preparing to join the WA. The list is periodically updated. SCOMET items are listed under nine (9) categories as follows:[15]

Category 0: Nuclear materials, nuclear-related other materials, equipment and technology. (The licensing authority for this category is the Department of Atomic Energy).

Category 1: Toxic chemical agents and other chemicals.

Category 2: Micro-organisms, toxins.

Category 3: Materials, Materials Processing Equipment, and related technologies.

Category 4: Nuclear-related other equipment, assemblies and components; test and production equipment; and related technology, not controlled under Category 0.

Category 5: Aerospace systems, equipment including production and test equipment, related technology, and specially designed components and accessories thereof.

Category 6: Munitions List.

Category 7: Electronics, computers, and information technology including information security.

Category 8: Special materials and related equipment, material processing, electronics, computers, telecommunications, information security, sensors and lasers, navigation and avionics, marine, aerospace, and propulsion.

The SCOMET list is a single integrated list of goods and technologies that could be used for the production, development, or use of WMD, their delivery systems, and other military munitions.

Licensing and Enforcement Procedures

Export controls are administered by DGFT who is also the principal licensing authority for a major part of the SCOMET list that inter alia contains dual-use items and technologies. 'DAE issues licenses for all the nuclear and nuclear dual-use commodities found in SCOMET Category 0.'[16] The export of these items is regulated under the Guidelines for Nuclear Transfers (Exports). Category-6 Munitions List items are subject to a No Objection Certificate (NOC) from the Department of Defence Production, Ministry of Defence. Export license applications are reviewed by an inter-ministerial working group chaired by the DGFT. According to Rajiv Nayan:[17]

> Any application for the supply of SCOMET items is cleared by an Inter-Ministerial Working Group (IMWG), with the Director General of Foreign Trade chairing meetings of the IMWG. Representatives from the Ministry of External Affairs, the Ministry of Defence, the Defence Research and Development Organisation, the Department of Defence Production, the Department of Atomic Energy, the Department of Space, the Indian Space Research Organisation, National Authority of the Chemical Weapons Convention, the Department of Chemicals, the Department of Chemical and Petrochemicals and intelligence agencies all participate in the IMWG.

Applications for dual-use items (mandatory online submission of export application, manual submission of original hard copy of End User Certificate is required) on the SCOMET lists are processed through an inter-ministerial working group for final decision of approval or denial. The IMWG, which routinely meets once a month, works on the basis of consensus. Export decisions by the DGFT are subject to a no objection from all members of the IMWG. The matter can be referred to the High Powered Committee headed by the DGFT in case there is no consensus and it could even be raised to the level of Cabinet Committee on Security. Certification requirements include end use/user assurance, non-diversion, non-replication/modification, information on the end-use location, and conditionality of no re-transfer without the permission of the Government of India. India also claims to be maintaining a Denied Entity List (DEL), which is a blacklist of entities who are barred from obtaining export licenses; however, this list is not public. A diagrammatic layout of the licensing process is given below:

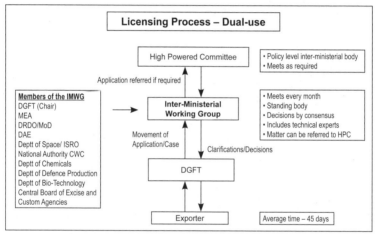

Source: Adapted from the Presentation by Anandi Venkateswaran, Under Secretary (D&ISA), Ministry of External Affairs and Kuldeep Parwal, Deputy Secretary (DIP), Department of Defence Production, India, given at the Wassenaar Arrangement Technically Focused Practical Workshop, held at Vienna on 27–28 June 2016.

The Indian government follows a multi-agency approach to enforcement. The FTDR Act provides for imposing financial and administrative penalties, and life imprisonment in case of major offences under the WMD Act. Maximum financial penalty is five times the value of goods being exported in contravention to the law. Interestingly, while punishment for offences under the WMD Act can extend to life imprisonment, under the Atomic Energy Act, which primarily deals with nuclear items, maximum liability is only five years imprisonment. Enforcement powers are vested in the Indian Customs. Authorised officials of Customs have the powers to search and seize, arrest, investigate, and impose financial penalties.

Supplier Capabilities

India ranks seventh in the world in terms of the number of operational reactors and fourteenth in terms of installed nuclear power capacity among the countries that are producing electricity through nuclear energy. India operates a closed nuclear fuel cycle programme. While nine power units are at different levels of construction, India claims to have achieved indigenisation and standardisation of 220 MWe pressurised heavy-water reactors. It has also exported 93.6 MT of heavy water during the period 2010–15.

As stated by Indian officials, India currently issues a few hundred export licenses of items on the SCOMET list; however, it is not certain how many of these export licenses items fall in the dual-use classification. Since Category-6 of the SCOMET list contains conventional munitions, it reinforces the widely held view that most of the export licensing activity would be in the realm of conventional weapons. India's effort to seek membership of the MECRs is motivated by the desire to gain greater access to advanced Western technology and, as the inflow of technology increases, so would the outflow of products; many of them could have dual uses. Current export of conventional munitions stands at $300 million annually. However, India has set its sight to enhance it to $2 billion in the next few years. This could only be possible if India gets access to sophisticated weapons technologies to boost its domestic industry. Following the Indo-US civil nuclear cooperation agreement and subsequent exemption from the NSG's full scope safeguards on nuclear transfers, India has signed bilateral cooperation agreements with over a dozen countries.

As declared by Indian officials, India issues 200 plus licenses per annum on the average covering exports to 30 countries, which include filament winding machines, UAVs, control valves, graphite heat exchangers, pumps, night vision devices, titanium pipes, CNC filament winding machines, and glass crystallised vessels etc.[18] Other export items include nuclear-grade graphite and heavy water. According to the June 2017 report of Project Alpha, there are 52 commercial dual-use items manufacturers in India.[19] Nonetheless, as India moves to integrate into international export control regimes and seeks more assistance from technologically advanced countries, its exports of dual-use items would steadily grow. India has already started to increase the range of its supersonic cruise missile Brahmos from the current 290 km to 450 km.[20] This has come after India joined the MTCR in June 2016. India has also stepped up efforts to accelerate its defence sales including Brahmos cruise missile, which many experts see as an outright violation of MTCR guidelines.

Engagement with International Export Control Regimes

The 2005 declaration by President Bush and Premier Manmohan Singh to transform the relationship between their respective countries and establish a global partnership provided the basis to gradually bring India into mainstream nuclear commerce. In early November 2010, President Obama visited India, and announced US support for India's full membership in the MECRs i.e. NSG, MTCR, Australia Group (AG), and Wassenaar Arrangement in a phased manner, hoping to elevate the nation of a billion people to 'its rightful place in the world' alongside an assertive China,[21] and pledged to remove some Indian entities from the US Department of Commerce's 'Entity List'. In 2012 Foreign Secretary Ranjan Mathai reiterated India's interest in joining the four export control regimes when he stated, 'While we wish to move forward in tandem on all the four regimes, our engagement with NSG is seen by observers as the most important. The logical conclusion of partnership with India is its full membership of the four multilateral regimes.'[22]

Consequently India has expanded its engagement with the MECRs and worked to align its export control policies and procedures with them therewith. The interest to seek membership of MECRs is motivated by the desire to enhance its prestige and international influence, acceptance as a nuclear weapon state, and to seek access to advanced technologies and equipment, and expand defence and high-technology procurement from foreign entities. India subsequently moved to declare adherence to the guidelines of MECRs, introduced new laws, revised the SCOMET list to incorporate MTCR, NSG, WA, and AG related items, and modified its licensing system to demonstrate greater alignment with international best practices on strategic trade controls. After joining the Hague Code of Conduct against Ballistic Missile Proliferation (HCOC), an informal and confidence building arrangement, India moved quickly to join MTCR (in June 2016), WA (in December 2017), and AG (in January 2018) whereas it has submitted its application for the NSG membership in May 2017. India's admission into the three MECRs has increased the prospects of its NSG membership as the regimes' members have cross-cutting linkages. This notwithstanding, India's participation in the NSG appears problematic in the face of strong opposition from China and other countries who emphasise on NPT as a fundamental requirement among the 'factors to be considered' for accepting new members. Currently NSG is at a critical juncture and grappling with the issue of participation of non-NPT states including India and Pakistan. For NSG to remain a viable non-proliferation arrangement, it must not override its principles for narrow commercial and political interests of certain partner countries.

Impact of the 2008 Exemption on Civil Nuclear Cooperation with India

The 2008 NSG exemption granted to India was a big blow to the international non-proliferation regime as it enabled India to clinch civil nuclear cooperation agreements with over a dozen countries and allowed it access to advanced nuclear technology. This has spared its indigenous resources for enhancing and modernising its nuclear weapons. Unabated fissile material production is leading to an arms race between India and China on the one hand, and India and Pakistan on the other, thus giving rise to destabilising tendencies in the region. It is understandable that in the changing geopolitical landscape of the region, India is being bolstered as a counter weight to China. But non-proliferation regimes should be more inclusive and should not create any exceptions. Discriminatory policies based on subordinating principles to politics will weaken international non-proliferation institutions and may fuel arms races.

The proliferation implications of the exceptions in the Indian safeguards agreement defy the spirit of NPT and IAEA safeguards, i.e. non-diversion of materials intended for peaceful uses to military purposes.[23] The separation plan agreed with the IAEA has not been fully implemented. Eight civilian power reactors, fast-breeder reactors, fuel cycle, and other facilities are still outside safeguards. Close links remain between military and civilian programmes, and scientific knowledge can be transferred from the civilian domain for enhancing and refining military capabilities. There is no change in India's opposition to the Comprehensive Test Ban Treaty (CTBT). Contrary to the spirit of the commitment of 2005, India has not ceased the production of fissile material for military purposes. In fact, huge new facilities for enrichment are being established in Karnataka and elsewhere outside the IAEA safeguards.[24]

Challenges to Strategic Export Controls

Over the past two decades, India has worked to reform its strategic export control system and has gradually aligned its policies and practices with MECRs to create a supportive environment for its membership. However, it still faces capacity issues. Though legislative and administrative measures have been developed, it is unknown how far these measures are effective since they have not been put to test. The public sector organisations may be following relatively strict monitoring of sensitive and dual-use items including checks over intangible technology. However, down the road, the expansion of the private sector will pose major problems for compliance with export control laws. Richard Cupitt says, 'It is not clear that India has adequate oversight of private firms working in sensitive dual-use areas, especially in the chemicals and pharmaceutical sectors.'[25] India has yet to introduce an internal compliance system and partnership with the exporting community in reinforcing compliance at the grass roots level. During the Asian Export Control Seminar at Tokyo in February 2018, an Indian representative from the DGTR office admitted that India has yet to work in this area. Comparing policy design, process, and implementation of non-proliferation export controls in five key countries, Richard, T. Cupitt and Scott A. Jones have found that 'while China, India and Russia have developed systems that approach international standards in terms of policy design and process, gaps remain in the area of implementation.'[26]

There are more than five hundred exit points in India including airports, seaports, and land borders that would require enormous resources for effective implementation of export control laws. Long and porous borders with other neighbouring countries lend easy opportunity for slippage of sensitive goods through unregulated areas; lack of technical capacity and suitable detection equipment accentuates implementation difficulties. Although the US has been supporting India in capacity building and enhancing competency in licensing, enforcement, and detection technologies and commodity identification, the latter has to overcome many difficulties. Incompatibility in the classification method between the SCOMET list and HS coding system of the Indian tariff code presents practical problems to Customs inspectors in identification of dual-use items at the export stage. Inherently, the Indian Customs is geared towards the scrutiny of imports,[27] which are an important revenue source; thus, there is relatively less focus on outbound goods.

'Make in India' slogan flags the shift towards market growth and potentially increased market share for private industry resulting from liberalisation of the economy, which is risk prone in case of the rapidly growing economies. Commenting on India's export control system, Ian Stewart opined:

> Whilst difficult to substantiate, it seems likely given the low number of license applications submitted to the Indian government that there is an amount of trade that takes place without the required licenses. These risks are compounded in India's case because of the country's isolation from the international nuclear market: Indian firms have indigenised capability of many nuclear components and technology.[28]

The chemical industry is a key constituent of India's economic and industrial growth, which is largely led by the private sector and is one of the most diversified industries. In terms of value and production volume, the Indian chemical industry is the third largest producer in Asia and sixth by output in the world.[29] Lack of awareness and lax compliance in the private sector presents enormous implementation challenges. This view is corroborated by the remarks of Dr Seema Gahlaut: 'the bulk of the foreign trading companies in the chemical sector are from the private sector, making the task of enforcement more challenging.'[30] Though India claims to have an unblemished non-proliferation record, there have been reports of Indian entities that provided chemicals to Iraqi entities; more so, Indian scientists have been working in Iran as consultants.[31] Transit and trans-shipment are the most neglected aspects in terms of enforcement and Customs jurisdiction. As concluded by Ritesh Kanodia and Aman Bhalla, 'while the legal authority and capability to deter, identify, and halt the transit and trans-shipment of specified goods, services, or technology have been put in place and are in effect, no specific guidelines or application procedures have been laid down for them.'[32]

India's non-proliferation record is far from 'impeccable' as there have been instances where Indian entities have directly or indirectly contributed to suspected military nuclear programmes. The growing involvement of the private sector in manufacturing dual-use and specially designed military goods enhances the risk of export control violations and circumvention of regulatory oversight. The proliferation case study of Project Alpha claims, 'India is a capable producer of dual-use goods and there have been instances where India-produced goods have reached programmes of proliferation concern.'[33] For example, many Indian entities have been involved in the transfer of proliferation-sensitive items and services to Iran and Iraq. Since 1992, unilateral US sanctions have been imposed on numerous Indian entities for transfer of proliferation-sensitive goods and technologies to countries of concern including Iran and Iraq, and sanctions were clamped on Hans Raj Shiv, Pro-tech Consultants Pvt Ltd, and NEC Engineers Pvt Limited for transferring WMD equipment, technology, and chemical and biological weapons materials.[34]

Conclusion

Historically, India had been critical of the international non-proliferation regimes and considered them discriminatory and a barrier to access to advanced technologies. However, after its economic liberalisation and subsequent rapprochement with the US to partner it against the rising Chinese power, India moved to align its strategic export controls' legislative, regulatory, and implementation mechanisms with MECRs though selectively accepting the non-proliferation obligations; for example, keeping eight power reactors and the fast-breeder programme outside the IAEA safeguards is contrary to international norms. Under the US pressure and support from UK, France, and other countries, India became a member of MTCR, WA, and AG while membership applications of some other countries are held up without any decision.

Incorporation of India into MECRs would provide it greater access to modern technologies and in turn enhance its manufacturing capacity of items of proliferation concern. India, through shrewd diplomacy, has been able to proceed selectively in engagement with the international non-proliferation regimes and has parried pressures to accept stricter restrictions on potential export of sensitive dual-use goods. Over the past three decades, India has considerably improved regulatory controls over the transfer of sensitive and dual-use items; however, serious gaps remain in capacity for implementation. Increased market share of private industry resulting from economic liberalisation is prone to risk in the case of the rapidly growing economies.

Notes and References

1. Munir Akram, 'India's Great Power Game', *Dawn*, 28 September 2014.
2. Statement by the US President George W. Bush, on 'Next Steps in Strategic Partnership with India', released by the White House Office of the Press Secretary, 12 January 2004 <https://2001-2009.state.gov/p/sca/rls/pr/28109.htm>
3. Brian Shoup, Sumit Ganguly, and Andrew Scobell, *US-Indian Strategic Cooperation into the 21st Century: More than Words* (Routledge, 2007), pp. 4–5.
4. Dr Rajiv Nayan and Ian J. Stewart, 'Export Control and India' (London, UK: Centre for Science and Security Studies, King's College London, August 2013).
5. Presentation by Devika Lal and Sanjit Kumar Samal (Under Secretary, Disarmament and International Security Affairs Division, Ministry of External Affairs, Government of India and Joint Director General of Foreign Trade, Ministry of Commerce and Industry, Government of India) at the 22nd Asian Export Control Seminar Held at Tokyo, Japan, 17–19 February 2015.
6. Ibid.

7. The Institute of Chartered Accountants of India, *Handbook on Foreign Trade Policy and Guide to Export and Import* (New Delhi: ICAI Bhawan, 2008).

8. Indian Atomic Energy Act 1962, Sections 14 and 16.

9. The Institute of Chartered Accountants of India, *Handbook on Foreign Trade Policy and Guide to Export and Import* (New Delhi: ICAI Bhawan, 2008).

10. India Directorate General of Foreign Trade, accessed at <http://www.dgft.org/>

11. Permanent Mission of India to the International Organisations in Vienna letter No. VIEN/110/19/2005, 17 June 2005 <https://www.iaea.org/sites/default/files/publications/documents/infcircs/2005/infcirc647.pdf>

12. Ibid.

13. 'Historical Background of Export Control Development in Selected Countries and Regions', available at <http://www.cistec.or.jp/english/service/report/1605historical_background_export_control_development.pdf>; Gaurav Rajen, 'Nuclear Export Controls in India: A Review of Existing Systems and Prospects for Enhanced India-US Cooperation', SASSI Research Report 12 (London: South Asian Strategic Stability Institute), April 2008.

14. Rajen, 'Nuclear Export Controls in India: A Review of Existing Systems and Prospects for Enhanced India-US Cooperation', April 2008.

15. Laxman K. Behera, 'Changes in the SCOMET List: What it Means for the Indian Defence Industry' <https://idsa.in/idsacomments/changes-in-the-scomet-list--indian-defence-industry_lkbehera-gbala_080617>.

16. Rajen, 'Nuclear Export Controls in India', April 2008.

17. Dr Rajiv Nayan and Ian J. Stewart, 'Export Control and India' (London, UK: Centre for Science and Security Studies, King's College London) August 2013.

18. Presentation by Anandi Venkateswaran, Under Secretary (D&ISA), Ministry of External Affairs and Kuldeep Parwal, Deputy Secretary(DIP), Department of Defence Production, India, given at the Wassenaar Arrangement Technically Focused Practical Workshop, Vienna, 27–28 June 2016.

19. 'India's Strategic Nuclear and Missile Programmes: A baseline Study for Non-proliferation Compliance', Project Alpha, Centre for Science and Security Studies, King's College London, June 2017.

20. Rahul Singh, 'From 290 km to 450 km: India to soon test extended range BrahMos missile,' *Hindustan Times*, 16 February 2017 <https://www.hindustantimes.com/india-news/brahmos-cruise-missile-range-to-increase-to-450km-drdo-chief/story-iJi2AKJBlEblDEyoxnLyKO.html>

21. 'Obama Endorses India's bid for Permanent Seat in the UNSC',

Times of India, 8 November 2010.

22. Institute of Defence Studies and Analyses, 'Full Membership for India in Export Control Regimes should be the Next Logical Step, Says Mathai,' Press Release, 18 April 2012 <http://idsa.in/pressrelease/FullMembershipforIndiainExportControlRegimes Mathai>

23. Kamran Akhtar, Director General (Disarmament), Ministry of Foreign Affairs, Islamabad, 'NSG Membership: Non-NPT States', *Islamabad Papers 2016*, Nuclear Paper Series No. 4 (Islamabad: Institute of Strategic Studies, August 2016).

24. Ibid.

25. Richard T. Cupitt and Scott A. Jones, 'Harmonization and Development of National Export Control Systems', in Michael D. Beck, et al. (eds.), *To Supply or To Deny: Comparing Non-Proliferation Export Controls in Five Key Countries* (The Hague: Kluwer Law International, 2003).

26. Ibid.

27. Presentation by Devika Lal and Sanjit Kumar Samal at the 22nd Asian Export Control Seminar Held at Tokyo, Japan, 17–19 February 2015.

28. Dr Rajiv Nayan and Ian J. Stewart, 'Export Control and India'.

29. IBEF, 'Chemicals,' May 2017 <https://www.ibef.org/download/Chemicals-May-2017.pdf>

30. Dr Seema Gahlaut, 'Indian Export Control Policy Political Commitment, Institutional Capacity, and Nonproliferation Record' (USA: Center for International Trade and Security [CITS], University of Georgia, January 2008).

31. Satinder Bindra and Amol Sharma, 'Probe into Illegal Indian Exports to Iraq', *CNN*, 26 January 2003; Bill Gertz, 'Indian Scientists Sanctioned for Assisting Iran on Nukes', *Washington Times*, 22 October 2004.

32. Ritesh Kanodia and Aman Bhalla, 'Transit and Transshipment of Dual Use Items', *WorldECR* (London: DC Houghton Ltd, 2012).

33. Proliferation Case Study Series, Project Alpha, 17 September 2014 <http://projectalpha.eu/proliferation-case-study-series-valves-for-arak>

34. Report of the US Senate Committee on Foreign Relations on US-India Peaceful Atomic Energy Cooperation and US Additional Protocol Implementation Act, 20 July 2006 <https://www.govinfo.gov/content/pkg/CRPT-109srpt288/html/CRPT-109srpt288.htm>

5

India's Nuclear Regulatory Regime

Happymon Jacob and Tanvi Kulkarni

Introduction

Even though it contributes modestly to the country's energy basket, India's civilian nuclear programme is expansive, one that the country has developed over the past several decades. With a current installed capacity of 6,780 MWe and twenty-two operational nuclear reactors,[1] nuclear power contributes about 3 per cent to India's total energy mix. As the demand for energy and electricity within the country grows, India is beginning an ambitious expansion of the country's nuclear power sector, with several nuclear projects being planned and in the pipeline, and more international engagements for nuclear commerce. In the aftermath of the Nuclear Suppliers Group (NSG) exemption to India and the Indo-US Nuclear Deal finalised in 2008, India's nuclear energy industry has evidently flourished. As per official records, India's nuclear power generation has recorded a growing trend since 2008.[2] In 2016, India recorded the highest ever annual nuclear power generation in its history.[3] India is also expected to undertake greater and more significant participation in international nuclear commerce.

The overall picture of India's nuclear energy programme, however, has not been rosy. The story of nuclear power generation in India has also been one of unachieved targets, technological, economic, and bureaucratic hurdles, and safety, security, and regulatory concerns. As India embarks on its ambitious nuclear energy expansion plans, addressing these concerns, especially those related to safety and security aspects of its nuclear programme, will be a serious challenge. Nuclear regulation, in particular, is an integral element determining the success of India's nuclear energy programme. India's nuclear regulatory framework, therefore, merits a meaningful discussion.

Even though nuclear power makes negligible contributions to India's overall energy basket, it remains a vital component of India's evolving energy security strategy and sustainable development plans. Given the projected growth of India's energy requirements—energy consumption is expected to grow at about 4.2 per cent per year by 2035,[4] faster than other major economies around the world—India has planned a large-scale expansion of its nuclear power industry and infrastructure. The Department of Atomic Energy (DAE) aims at installing 20,000 MWe of power generation capacity by the year 2020.[5] An even more ambitious target set by the department is to cover 25 per cent of India's total electricity generation with nuclear power by 2050.[6]

India is also currently in the second stage of its planned three-stage nuclear programme, with nuclear power being generated through the first-stage pressurised heavy-water reactors (PHWR) and second-stage fast breeder reactors (FBR). The next step of the three-stage programme will consist of the thorium fuel cycle. The Indian nuclear power programme faces several regulatory challenges, particularly with respect to autonomy and effectiveness. These criticisms become rather severe in the face of nuclear incidents and accidents. Unless these challenges are addressed satisfactorily, the next stage of the nuclear power programme is expected to bring further financial, technical, and regulatory challenges.

This chapter looks at the historical and institutional evolution of civilian nuclear regulation in India in the broader context of the international nuclear regulatory framework. The chapter does not discuss the regulation of non-civilian nuclear activities in the country nor does it focus on nuclear security; the focus is exclusively on nuclear regulation in the civilian sector. Section one of the chapter discusses the international nuclear regulatory framework to give a broad overview of the global context within which the Indian civilian nuclear regime functions. Section two examines India's nuclear regulatory framework, its evolution, and current structure. This section will examine how major events, incidents, and instruments—in the international experience as well as the Indian domestic experience— have shaped the Indian discourse and nuclear regulatory setup. The final section of the chapter looks at the major concerns pertaining to the nuclear regulatory framework in the country, and in doing so, discusses various recommendations and suggestions that seek to address those concerns.

Understanding Nuclear Regulation

Although the concept of an international regulatory regime challenges the traditional notion of Westphalian sovereignty, what exists today in the case of international nuclear regulation is a regime which primarily depends on state agencies and actors for the implementation of regulatory functions. This arrangement appears to create an overlap, and even some confusion, about the boundary between international regulation and domestic regulation.[7] While the global nuclear regulatory regime—also called the Global Nuclear Safety Regime—is a framework which seeks to implement the highest standards of safety at nuclear installations around the world,[8] at the core of the regime are the national commitments and state-led efforts at ensuring and enhancing nuclear safety. Nuclear safety and security are ultimately the responsibility of states and the regulation of their nuclear programmes is done through domestic institutions constituted for the purpose.

The International Atomic Energy Agency (IAEA) defines a regulatory authority (superseded by the term 'regulatory body') as 'an authority or authorities designated or otherwise recognised by a government for regulatory purposes in connection with protection and safety (of nuclear and radioactive materials).[9] The IAEA Safety Glossary explains:

> The definition of *regulatory body* indicates the conditions that must be met in order for an organisation to be described as a *regulatory body*, but not the attributes of a *regulatory body* as required by IAEA safety standards. Hence, the definition specifies that it is 'designated by the government of a State as having legal authority for conducting the regulatory process'—otherwise, it is not a *regulatory body*. However, the definition does not, for example, specify that it is 'independent of organisations or bodies charged with the promotion of nuclear technologies'—it can be a *regulatory body* without being independent, even though it would then not satisfy the IAEA Safety Requirements on legal and governmental infrastructure for safety.[10]

Regulatory control is defined by the IAEA as 'any form of control or regulation applied to facilities or activities by a regulatory body for reasons relating to radiation protection or to the safety or security of radioactive sources.'[11] Further, the agency identifies three approaches to nuclear regulation:[12]

a) Compliance-based regulation, where operators have to follow prescriptive standards and requirements, and inspection and enforcement are largely a matter of verifying compliance with rules and penalising non-compliance.

b) Performance-based regulation, where the regulator uses safety performance indicators and inspections to assess safety. The licensees have some flexibility to decide how to achieve prescribed safety objectives. This approach involves the risks of manipulation of indicators by licensees.

c) Process-based regulation (or integral supervision of nuclear power plants), which recognises the flexibility of the design of organisational processes to create processes that are internally consistent, adapted to their history, culture, and business strategy, and that allocate resources in the most rational way.

The three approaches are not mutually exclusive and can be employed as an integrated approach to establish and ensure a good nuclear safety culture at the national and international levels.

Defining Nuclear Regulation, Nuclear Safety and Security Culture

Let's briefly look at some of the standard definitions of nuclear regulation, nuclear safety, and security culture to understand as well as to differentiate among them. The IAEA defines nuclear safety culture as the 'assembly of characteristics and attitudes in organisations and individuals which establishes that, as an overriding priority, protection and safety issues receive the attention warranted by their significance.'[13] It defines nuclear security culture as the 'assembly of characteristics, attitudes and behaviour of individuals, organisations, and institutions which serve to support and enhance nuclear security.'[14] An effective nuclear regulatory framework has to address the shared objectives of the safety and security cultures: 'to limit the risks resulting from radioactive material and associated facilities.'[15] In its narrowest sense (nuclear) regulation can be defined as the 'promulgation of an authoritative set of rules, accompanied by some mechanism, typically a public agency, for monitoring and promoting compliance with these rules.'[16] In the IAEA Nuclear Safety Glossary, nuclear regulation has been defined through the word 'control' meaning 'the function or power or (usually as controls) means of directing, regulating or restraining.'[17]

The International Nuclear Regulatory Regime

Evolution

The International Nuclear Regulatory Regime is the institutional, legal, and technical framework for ensuring the safety and security of civilian nuclear installations around the world. Although nuclear regulation is primarily associated with nuclear safety, international nuclear regulation is also closely concerned with nuclear security and safeguards of radioactive and nuclear facilities and materials.

The Global Nuclear Safety Regime and the international nuclear regulatory framework came up because of the worldwide expansion of nuclear power programmes in the 1960s and 1970s. Apart from the development of new types of nuclear reactors and facilities,[18] designs, approaches, and the increasing number of operators and vendors, there was also a growth in international commerce in the nuclear power industry. New discoveries came up in nuclear technology and countries developed national regulatory systems at the domestic level to manage their atomic power programmes. The fears of nuclear technology being misused led to the creation of the Atoms for Peace programme, first expressed as an idea by US President Eisenhower in 1953. The IAEA, as the successor organisation to the Atoms for Peace initiative, was established on 29 July 1957.[19]

Apart from major transformations in the political, economic, and technological domains, periods of crises also led to a critical scrutiny of existing conditions.[20] The urgency to bring countries into a single Global Nuclear Safety Regime was felt after the accident at Chernobyl in 1986.[21] The Convention on Nuclear Safety (CNS) was signed in 1994 and came into force in 1996. Many international conventions for nuclear safety and security have been signed thereafter for the joint development of safety regulations and the establishment of international networks among nuclear power plant operators and national regulators.[22] While the 9/11 attacks in New York spurred a worldwide focus on nuclear security, taken up internationally through the Nuclear Security Summits, the Fukushima Daiichi incident in March 2011 renewed the debate about nuclear safety (and security). The assessments and lessons learnt from these incidents have impacted nuclear regulatory practices and approaches at the national/domestic and international levels.

In 2006, the IAEA held its first international conference on 'Effective Nuclear Regulatory Systems Facing Safety and Security Challenges' in Moscow. The conference concluded that the key challenges facing regulatory authorities and nuclear industry come from: the expansion of nuclear industry with a renewed interest in nuclear energy; the increased security threats to nuclear installations; increased global use of radioactive materials; new safety and security challenges from old nuclear installations like ageing; and extension of operating lifetimes.[23] Over a decade later, the challenges remain similar in addition to new ones. Ten years later, the 2016 International Conference on Effective Nuclear Regulatory Systems held in Vienna identified four areas of challenges: diversity of reactor designs, radiation sources and radioactive waste, strengthening international cooperation, and strengthening regulatory competence in terms of adequate technical, managerial, and human resources.[24]

Structure

As of October 2017, there are 448 nuclear reactors generating electricity in thirty countries around the world, representing a total net installed capacity of 391,744 MWe.[25] About fifty-seven new nuclear plants are under construction[26] in fifteen countries. Nuclear power plants provided 11 per cent of the world's electricity production in 2014.[27]

The IAEA, headquartered in Vienna and part of the United Nations family, is the institutional hub for nuclear safety and security, science and technology, and safeguards and verification. It is important to note that the IAEA is not an international nuclear regulatory body, since it does not have the supranational authority to regulate the activities and behaviour of national nuclear energy programmes. However, the IAEA performs important regulatory functions: it acts as the secretariat for all safety-related conventions; develops and promotes safety standards, guidelines, and codes of conduct; manages safety advisory missions and peer review processes; promotes the peaceful uses of nuclear energy and assists member-states in planning and using nuclear science and technology for various peaceful purposes, including electricity generation; and verifies compliance to peaceful uses of nuclear energy through its inspection system.

The CNS is the most important legal instrument of the International Nuclear Regulatory Regime. The convention is an international treaty under the IAEA which legally binds its member-parties, currently at eighty signatories, to ensure high standards of nuclear safety through national measures and international cooperation.[28] The scope of the CNS is limited only to land-based civilian nuclear installations and it does not extend to nuclear installations at sea, in space, or to military nuclear facilities. The CNS does not have monitoring systems to ensure compliance; this raises questions about transparency and accountability.[29] The CNS requires states-parties to establish legislative and regulatory bodies with adequate authority, competence, and financial and human resources to fulfil its assigned responsibilities and to maintain an effective separation of functions between those of the regulatory body and other organisations involved in the promotion and utilisation of nuclear energy.[30] The CNS also mandates contracting parties to undertake verification (Article 14) and quality assurance (Article 13) through inspection, testing, analysis, surveillance, and assessments.

Apart from the IAEA and the CNS, the International Nuclear Regulatory Regime is also composed of several other intergovernmental organisations, such as the Organisation for Economic Co-operation and Development/Nuclear Energy Agency (NEA); multinational networks of nuclear regulators like the International Nuclear Regulators Association (INRA), Network of Regulators of Countries with Small Nuclear Programmes(NERS), Western European Nuclear Regulators Association (WENRA), European Nuclear Safety Regulators Group (ENSRG), etc.; legal instruments like the Convention on Early Notification of a Nuclear Accident (1986), Convention on Assistance in the Case of Nuclear Accident of Radiological Emergency (1987), Convention on the Physical Protection of Nuclear Material (CPPNM) (1987), Joint Convention on the Safety of Spent Fuel Management and on the Safety of Radioactive Waste Management (2001), and Additional Protocol to the CPPNM (2005); codes of conduct and guidelines by the IAEA and safety review processes like the Integrated Regulatory Review Service (IRRS) and Operational Safety Review Team (OSART). The International Nuclear Regulatory Regime together with the multinational network of operators like the World Association of Nuclear Operators (WANO) or the International Network for Safety Assurance of Fuel Manufacturers (INSAF),

Fig 1: Main elements of the Global Nuclear Safety Regime

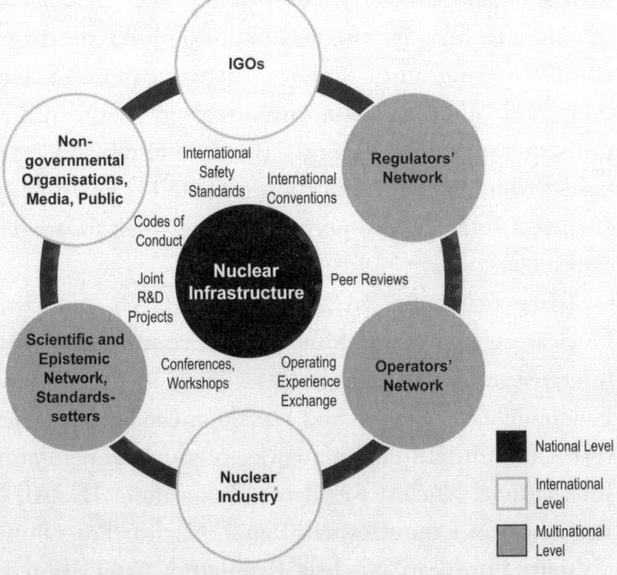

Source: Adapted from the INSAG–21A Report by the International Nuclear Safety Group, IAEA <http://www-pub.iaea.org/MTCD/Publications/PDF/Pub1277_web.pdf>

national nuclear infrastructures, and other stakeholders including suppliers of nuclear fuel, equipment, and other services, vendors, non-government organisations (NGOs), and the nuclear liability regime through which the operators of nuclear power plants (NPPs) can be held accountable for nuclear accidents or incidents through legal proceedings and liability payments, make up the Global Nuclear Safety Regime.

India's Nuclear Regulatory Regime/ Framework

India and the Global Nuclear Order

Before we examine India's nuclear regulatory framework, it would be useful to briefly situate India's nuclear journey and its relationship with the global nuclear order in proper historical context.

India's current engagement with the international nuclear order can be described as a second coming, after India's period of active opposition to the order and the NPT from the mid-1950s to the 1980s. By the late 1980s, India realised it was fighting a losing battle and became serious about nuclear weapons—even while it advocated the Rajiv Gandhi Action Plan for Nuclear Disarmament.[31] India has a remarkable history of anti-nuclear activism, proposing an end to nuclear testing in 1954 after the US nuclear testing in Bikini Atoll,[32] and signing the Partial Nuclear Test Ban Treaty (PTBT) in 1963. It played a major role in the negotiations to establish the IAEA and actively participated in the negotiations on the NPT, but decided not to sign it eventually.

Indeed six of its nuclear reactors were voluntarily made subject to IAEA inspections even though it was under no obligation to do so, not being a party to the NPT. Today, India is open to negotiating a Fissile Material Cut-off Treaty (FMCT) and is no longer as opposed to signing the Comprehensive Nuclear-Test-Ban Treaty (CTBT)[33] as it was in the mid-1990s.[34] India was one of the co-sponsors of the CTBT resolution at the 1993 session of the UN General Assembly.

India eventually did not sign the CTBT for a variety of reasons, the most important being the treaty's inbuilt discrimination.

New Delhi's major objection is to the NPT. The NPT does not recognise India as a nuclear-weapon state which the latter insists it is. The then Foreign Minister, Jaswant Singh, articulated India's position on the NPT in 2000: 'India's policies have been consistent with the key provisions of NPT that apply to weapon states. India has been a responsible member of the non-proliferation regime and will continue to take initiatives to bring about stable and lasting non-proliferation.'[35] However, India has also clarified that it will never join the NPT as a non-nuclear-weapon state.

In the mid-1990s the international community, led by the United States, was calling on India to roll back its nuclear weapons programme[36] and sign the CTBT that was then being negotiated. When India tested its weapons in the summer of 1998, the international community responded with condemnation and sanctions.[37] This, however, did not last too long. The international nuclear order showed willingness to negotiate with India.

After years of painstaking negotiations, India and the United States announced an Indo-US nuclear deal in 2005 and eventually signed it in October 2008. The deal brought India straight into the centre of the global nuclear order. The bargain the two countries had struck was mutually beneficial; New Delhi did not have to give up its nuclear weapons to be part of the international nuclear order and the NPT did not have to be rewritten to accommodate India. The IAEA, after the US Congress passed the deal, negotiated an India-specific protocol in 2009 so as to place India's civilian reactors under its inspection regime.[38] India also separated its civilian and military reactors so that there is no cross-feeding of nuclear material from one to the other. Whatever nuclear material India received from nuclear-supplier countries would, of course, be used for civilian purposes.

In 2008, after considerable negotiations, the NSG, which normally prohibits its members from nuclear commerce with states which have not signed the NPT, agreed on a special waiver in the case of India.[39] However, it continues to be unclear whether India will be able to benefit from enrichment and reprocessing (ENR) technologies, about the transfer of which the NSG is framing new rules.[40] Although in amending its guidelines in 2011, NSG stressed it would restrict ENR commerce to parties to the NPT, the United States, France and Russia have said that they will continue with the clean exemption the NSG gave India in 2008.[41] Even so, for purely commercial reasons it is unclear how willing they will be to transfer ENR technology to India. Since the NSG waiver, New Delhi has signed nuclear deals with several countries. Today, India is a member of the Missile Technology Control Regime, Wassenaar Arrangement, and Australia Group and is campaigning to formally join the Nuclear Suppliers Group.

India and the International Nuclear Regulatory Regime

India was an early recipient of foreign aid for its nuclear power programme under the Atoms for Peace programme. Even during the phase of self-regulation, India submitted a design and safety report for the Canada India Reactor Utility Services (CIRUS) reactor to the Canadians at their insistence.[42] While India was under three decades of international nuclear sanctions after 1974, Indian NPPs tried to keep up with international benchmarks and best practices for design and operations while pursuing an indigenous nuclear programme. In the evolution of the safety, security and regulatory regime, the Indian regulatory body has aimed at incorporating the best practices existing in the international milieu. In fact, a cursory glance at the Atomic Energy Regulatory Board's (AERB) safety-related publications would show how they are influenced by IAEA's own publications.

As a result of an expansion and growth of India's nuclear commerce since the deal, India has been encouraged to integrate further into the global nuclear regulatory framework and regime. Most significantly, the immediate regulatory consequence of India's venture into the international nuclear commerce following the deal, was the civil nuclear liability legislation. The Indian civil nuclear liability act of 2010 currently puts heavy onus and severe liability over the suppliers of nuclear fuel and other material to Indian NPPs unlike international liability practices. New Delhi has since tried to address this problem.

In March 2015, for the first time the Government of India, on the request of the AERB, invited the CNS' Integrated Regulatory Review Mission (IRRS) to review India's nuclear regulatory framework for NPPs. The IRRS found that India has a sophisticated and well-developed nuclear regulatory regime with an experienced, knowledgeable, and dedicated regulatory body[43] but one which is still evolving in terms of best practices and areas that need further attention. The IRRS team identified the following good practices:[44]

> India has a comprehensive and well established national educational and training system that supports competence building for its nuclear programme.
> The AERB takes full benefit from operational experience with the aim of continuously enhancing its regulatory framework and processes.
> The AERB's research and development infrastructure provides strong regulatory review and assessment activities.
> The scope and depth of the AERB recruitment and training programme is effective in maintaining a knowledgeable technical staff.

The IRRS review team identified the following areas that need attention:[45]

> The Indian Government should promulgate a national policy and strategy for safety, as well as a radioactive waste management strategy as a statement of the Government's intent.
> The Government should embed in law the AERB as an independent regulatory body separated from other entities having responsibilities or interests that could unduly influence its decision making.

The AERB should review the implementation of its policy and existing arrangements to ensure it maintains independence in the performance of its regulatory functions.

The AERB should consider increasing the frequency of routine on-site inspections at NPPs. The increased frequency of inspections would allow for additional independent verification and more effective regulatory oversight of NPPs.   •

The AERB should develop and implement its own internal emergency.

Currently, India has nuclear commercial agreements with almost eight countries around the world and is negotiating business with almost a dozen more, including Germany, South Korea, Niger, Uzbekistan and Brazil. This expansion, diversification, and privatisation in India's nuclear energy will throw up challenges for India's regulatory regime including synchronising regulations pertaining to commerce with players from different countries.

From Self-Regulation to Regulatory Body

India's nuclear power programme started in the 1950s without a formal regulatory system. During this phase of self-regulation, the safety and design reviews at India's first research reactor at Trombay, the Apsara, were personally carried out by Homi Bhabha, who was the founding director of the Atomic Energy Establishment at Trombay.[46] International cooperation in the subsequent years however necessitated structured safety reviews and, as a result, a formal reactor safety committee was set up in 1962 with A. S. Rao as chairman for the operation of the Apsara, commissioned in 1956, and the CIRUS research reactor, commissioned in 1960.[47] It was not until February 1972 that a dedicated organisation and personnel for conducting the safety review—Safety Review Committee (SRC)—was constituted within the DAE to maintain safety oversight in its nuclear installations.[48] The SRC, it may be noted, was an internal review committee and not an independent regulatory body.

The seven decades of India's nuclear energy programme include an almost thirty-year phase of international technology sanctions after the country's first nuclear test in 1974 and further proscription from the international nuclear industry and commerce after India's refusal to join the Nuclear NPT. In the aftermath of sanctions, during which India's nuclear programme was built mostly indigenously, India's nuclear programme witnessed rapid progress in terms of designing, developing, and manufacturing equipment for the PHWRs and subsequently for India's FBRs.[49] It was during this phase that the AERB was set up as the country's apex regulatory body to oversee and enforce safety in all nuclear operations, including those within the DAE as well as those among the national industrial and medical users of radiation technology.[50]

While the safety regulation under the DAE created a well-integrated review, monitoring, and surveillance system, they were all controlled by the DAE. This model that seeks to prioritise safety over strict independence of the regulatory body appears to have later translated to the functioning of the AERB which was created in 1983. The AERB's regulatory practices have benefited from the experience and technological support that it gained through the operational safety reviews of the various BARC facilities between 1983 and 2000. In April 2000, following the declaration of BARC as a nuclear weapons laboratory, all of its facilities were excluded from the AERB's jurisdiction.

Legislative Framework for Civilian Nuclear Regulation

The overarching source of AERB's nuclear regulatory powers is the Atomic Energy Act of 1962 even though the AERB was set up much later in 1983. The government has over the years brought about several statutory provisions for issuing regulatory consent for nuclear facilities: Atomic Energy (Radiation Protection) Rules, 2004; the Atomic Energy (Working of the Mines Minerals and Handling of the

Prescribed Substances) Rules, 1984; the Atomic Energy (Safe Disposal of Radioactive Wastes) Rules, 1987; the Atomic Energy (Factories) Rules, 1996; and the Atomic Energy (Radiation Processing of Food and Allied Products) Rules, 2012. The AERB also draws some of its functional authority from the Environment Protection Act of 1986 and the Environment Protection Rules, 1987. The Civil Nuclear Liability for Nuclear Damage Act 2010 and the Civil Liability for Nuclear Damages Rules 2011 mandate the AERB to notify and give wide publicity to the occurrence of all significant nuclear incidents (Clause 3). The AERB is also governed under the Right to Information Act of 2005.[51] These legislations together form the legal foundation of the country's nuclear and radiation safety policies.

The Regulatory Body

The apex nuclear regulatory body in India, which looks after the safe and secure development of the nuclear energy sector, is the AERB. It is the national authority designated by the Government of India with the legal authority to issue regulatory consent for various activities related to nuclear[52] and radiation facilities,[53] and to perform safety and regulatory functions including the enforcement for the protection of site personnel, the public, and the environment against undue radiation hazards.[54] The primary function of the AERB is to assess and grant consent to nuclear and radiation-related activities, in the form of authorisations or licences. The AERB ensures that the activities of the nuclear/radiation facility comply with the safety requirements and conditions of consent.[55] This is done by the AERB through its system of safety reviews, assessments, and regular inspections. The safety and security objectives to be met by the consentees are issued through the AERB's safety codes, standards, and guidelines which are published through its safety documents and manuals. Consentees are mandated to develop quality assurance programmes specifying their goals, strategies, and plans, including individual and

Fig 2: Regulatory documents pursuant to primary legislation pertaining to nuclear energy

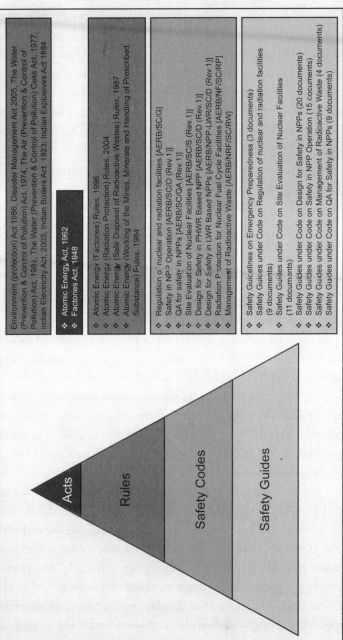

Acts

❖ Environment (protection) Act 1986, Disaster Management Act 2005, The Water (Prevention & Control of Pollution) Act, 1974, The Air (Prevention & Control of Pollution) Act, 1981, The Water (Prevention & Control of Pollution) Cess Act 1977, Indian Electricity Act, 2003, Indian Boilers Act, 1923, Indian Explosives Act 1884
❖ Atomic Energy Act, 1962
❖ Factories Act, 1948

Rules

❖ Atomic Energy (Factories) Rules, 1996
❖ Atomic Energy (Radiation Protection) Rules, 2004
❖ Atomic Energy (Safe Disposal of Radioactive Wastes) Rules, 1987
❖ Atomic Energy (Working of the Mines, Minerals and Handling of Prescribed Substance) Rules, 1984

Safety Codes

❖ Regulation o nuclear and radiation facilities [AERB/SC/G]
❖ Safety in NPP Operation [AERB/SC/O (Rev.1)]
❖ QA for safety in NPPs [AERB/SC/QA (Rev.1)]
❖ Site Evaluation of Nuclear Facilities [AERB/SC/S (Rev.1)]
❖ Design for Safety in PHWR Based NPP [AERB/SC/D (Rev.1)]
❖ Design for Safety in LWR Based NPPs [AERB/NPP-LWR/SC/D (Rev.1)]
❖ Radiation Protection for Nuclear Fuel Cycle Facilities [AERB/NF/SC/RP]
❖ Management of Radioactive Waste (AERB/NRF/SC/RW]

Safety Guides

❖ Safety Guidelines on Emergency Preparedness (3 documents)
❖ Safety Guides under Code on Regulation of nuclear and radiation facilities (9 documents)
❖ Safety Guides under Code on Site Evaluation of Nuclear Facilities (11 documents)
❖ Safety Guides under Code on Design for Safety in NPPs (20 documents)
❖ Safety Guides under Code on Safety in NPP Operation (15 documents)
❖ Safety Guides under Code on Management of Radioactive Waste (4 documents)
❖ Safety Guides under Code on QA for Safety in NPPs (9 documents)

Source: National Report to the Convention on Nuclear Safety, Seventh Review Meeting of Contracting Parties, March 2017, Government of India, p. 39.

organisational responsibilities, as well as to make emergency response plans.[56] The AERB has adopted a formal code of ethics comprising the fundamental principles and core values to be followed by its employees in the conduct of their duties and responsibilities.[57] The functions of the AERB can be summarised as follows:[58]

1. Develop safety policies in nuclear, radiological, and industrial safety areas.

2. Develop Safety Codes, Guides, and Standards for siting, design, construction, commissioning, operation, and decommissioning of different types of nuclear and radiation facilities.

3. Grant consents for siting, construction commissioning, operation, and decommissioning after an appropriate safety review and assessment, for establishment of nuclear and radiation facilities.

4. Ensure compliance of the regulatory requirements prescribed by AERB during all stages of consenting through a system of review and assessment, regulatory inspection, and enforcement.

5. Prescribe the acceptance limits of radiation exposure to occupational workers and members of the public and approve acceptable limits of environmental releases of radioactive substances.

6. Review the emergency preparedness plans for nuclear and radiation facilities and during transport of radioactive sources, irradiated fuel, and fissile material.

7. Review the training programme, qualifications, and licensing policies for personnel of nuclear and radiation facilities and prescribe the syllabi for training of personnel in safety aspects at all levels. Assessment of competence of key personnel for operation of NPPs.

8. Take such steps as necessary to keep the public informed on major issues of radiological safety significance.

9. Promote research and development efforts in the areas of safety.

10. Maintain liaison with statutory bodies in the country as well as abroad regarding safety matters.

11. Review of 'Nuclear Security affecting Safety' at nuclear installations.

12. Notify nuclear incident under Civil Liability for Nuclear Damage Act, 2010.

The AERB has seven technical divisions, and a safety research institute (SRI) located at Kalpakkam, Tamil Nadu. It has two safety review committees (Safety Review Committee for Operating Plants—SARCOP, and Safety Review Committee for Applications of Radiation—SARCAR), and an advisory committee (for Project Safety Review—ACPSR—comprising of field experts). The AERB adopts a three-tier safety review system for its regulatory consenting process— related to the major stages of siting, construction, commissioning, operation, and decommissioning of nuclear and radiation facilities— and a two-tier system for less hazardous facilities. Consentees submit detailed site evaluation reports (SER), safety analysis reports (SAR), and technical specifications for operating facility to the AERB's technical divisions, which are then referred to the safety review committee, which also gets input from the body's regulatory inspection reports. The AERB Board takes the final decision on issuing consent for setting up nuclear facilities.

According to the AERB policy, the regulatory standards and safety requirements have to be consistent with the IAEA standards on 'Governmental, Legal, and Regulatory Framework for Safety' and 'Fundamental Safety Principles'. The Indian nuclear regulatory framework interacts and cooperates with the international framework through various forums and treaty mechanisms. India is a member of several major safety, security, and safeguards-related organisations and treaties/ conventions, including the IAEA (joined in 1957), CNS (signed in 1994 and ratified on 31 March 2005), Convention on the Early Notification of Nuclear Accidents (signed in 1986 and ratified in 1988), Convention on Assistance in the Case of a Nuclear Accident or Radiological Emergency (signed in 1986 and ratified in 1988), CPPNM (acceded in 2002) and Amendment to the CPPNM (acceded in 2007), Convention on Supplementary Compensation for Nuclear Damage (signed in 2010 and ratified in 2016), and the Regional Co-operative Agreement for Research, Development and Training Related to Nuclear Science and Technology for Asia and the Pacific (joined in 2017).[59] AERB also participates in many multinational regulatory forums like WANO, CANDU Senior Regulators Group and the International Nuclear Events Scale (INES).

AERB, however, is not an autonomous regulator but a subordinate authority to the Government of India. The Atomic Energy Commission (AEC) of India is the apex governing body which frames the policies for the country's civilian nuclear programme. It was first constituted in 1948 under the Department of Scientific Research, and subsequently moved into the DAE by a government resolution in 1958. The DAE, established in 1954, is responsible for execution of policies laid down by the AEC and is engaged in research, technology development, and commercial operations in the areas of nuclear energy, and related high-end technologies. It supports basic research in nuclear science and technology.

Finally, the DAE has set up twenty-three emergency response centres (ERCs) at various nuclear power plants and radiation facilities with a nodal ERC at BARC.[60] The ERCs are equipped with trained emergency response teams (ERTs) as well as radiation detection instruments and systems for detection and response to nuclear/ radiological emergencies in the public domain. The operator of the NPP is responsible for the safety at the plant/ facility. Every NPP has a health physics unit (HPU), with trained and experienced radiation protection professionals, to implement the radiation protection programme in the plant.

Operators and Vendors

The Nuclear Power Corporation of India, Limited (NPCIL) carries out all activities related to siting, design, construction, commissioning, and operation of all nuclear power reactors in India. The NPCIL presently operates all of India's twenty-one nuclear power plants (including RAPS–1, which is under long shutdown and belongs to DAE).[61] The NPCIL is a Government of India organisation. The

Bharatiya Nabhikiya Vidyut Nigam Limited (BHAVINI) is yet another government-owned company that was established for the construction, commissioning, and operation of the fast breeder reactors.[62] The BARC and the Indira Gandhi Centre for Atomic Research (IGCAR) are the premier multi-disciplinary nuclear research centres for advanced research and development (R&D) in nuclear science and engineering. Other R&D centres are the Raja Ramanna Centre for Advanced Technology, Variable Energy Cyclotron Centre, Atomic Minerals Directorate for Exploration and Research, and the Global Centre for Nuclear Energy Partnership. The latter was set up in 2010 for conducting R&D and training in 'safe, secure, proliferation-resistant, and sustainable' nuclear systems[63] and is still to be fully operational.

A number of public sector units under the DAE manage and operate the various requirements of the nuclear fuel cycle and facilities. The Uranium Corporation of India Limited (UCIL) is responsible for mining, milling, and processing of uranium ore for the PHWRs, and the Nuclear Fuel Complex (NFC), another DAE unit, caters to the fuel fabrication and zirconium requirements of the Indian NPPs. The Electronics Corporation of India Limited (ECIL) designs, develops, and manufactures the electronic systems including computers, control systems, and communications. The Indian Rare Earths Limited (IREL) is responsible for production and sale of heavy minerals including monazite. A Nuclear Recycle Board has been established under DAE for developing and operating back-end fuel cycle facilities and for the reprocessing of spent fuel.

Increasingly, private sector companies with specialised domain knowledge in heavy engineering and manufacturing facilities have been participating in India's nuclear power industry. For instance, Larsen and Turbo provides heavy equipment like steam generators for India's NPPs.[64] Besides expansion, the Indian nuclear power sector also envisages a diversification plan for operational and management responsibilities pertaining to NPPs.[65]

The regulatory framework managed through the AERB's regulatory systems is only applicable to India's civilian nuclear facilities and NPPs. India's military nuclear programme is not subject to the AERB's regulatory processes. The AERB was previously associated with the licensing of plant personnel in all the critical installations. In June 2000, however, the regulatory and safety review functions related to BARC were transferred from AERB to an internal safety committee structure of BARC.[66]

Independence and Transparency

The issue of independence and transparency of India's nuclear regulatory system has been a major challenge vis-à-vis India's nuclear programme. AERB, India's apex nuclear regulatory body, is not an independent authority. The Atomic Energy Act of 1962, from where the Indian government derives the authority to manage and direct the Indian nuclear energy programme, clearly calls for the delegation of power by the central government to a 'subordinate authority'[67] (Article 27). The AERB is administratively responsible and accountable to, and financially dependent on, the AEC which also exercises overall governance of the nuclear industry through the DAE's organisational framework. The DAE is the promoter of nuclear energy in India and oversees the functioning of the operators—namely, the NPCIL and BHAVINI. While the AERB's jurisdiction does not extend to India's military nuclear facilities and it has renounced regulatory oversight over BARC since 2000, the apex regulatory body is dependent on BARC for staff, equipment, and technology like the BARC Channel Inspection System 2000 (BARCIS). Those who support the current organisational structure argue that the AERB enjoys 'functional autonomy' and that there have been no proven incidences where the AERB's functional autonomy has been compromised.

However, studies by experts have shown that there have indeed been safety compromises in India's NPPs and other facilities as a result of the lack of regulatory independence. These have come to light after significant accidents that took place in India's nuclear facilities. As pointed out earlier, the DAE's own evaluations of NPPs from 1979 and 1987 remained unaddressed for over a decade. After the 2003 incident involving a valve failure at the Kalpakkam Atomic Reprocessing Plant (KARP), a safety review committee advised the plant to be temporarily shut down. BARC, which holds administrative control over the facility, continued to operate it despite the committee's recommendations, leading to protests by the workers' union.[68] Moreover, quite apart from the issue of accountability, the risk of manipulation of processes and safety standards related to the safety of the Indian NPPs and radiation facilities by the DAE remains a serious issue for India's nuclear regulatory regime. Moreover, having a non-independent regulatory authority does not comply with benchmarks set by multinational forums like the CNS.

At least five significant reports/audits have pointed out the lacunae in the DAE's functioning and the serious dangers associated with the absence of a de jure independent nuclear regulatory authority: a report of an SRC constituted to look at the terms of reference of the DAE in 1981,[69] the 1997 Raja Ramanna Committee report which recommended amendment to the AEA (1962) to enhance the effectiveness of India's nuclear regulatory system,[70] the report of the Indian Comptroller and Auditor General (CAG)'s Performance Audit on the Activities of the AERB on 23 August 2012, the Indian parliament's Public Accounts Committee's (PAC) scrutiny[71] of the CAG report in 2013, and the IRRS report in 2015. These reports/audits pointed out the need for a clear legal and regulatory separation between the functioning of the AERB and authorities like the AEC and DAE, the provisions for grievances redress system in India with operators, suppliers, and vendors and other actors in the regulatory framework, including personnel, and greater clarity in the AERB's constitution.

Responding to the recurrent criticism and recommendations from across these high-level audit authorities, and necessitated by the Fukushima Daiichi accident, the United Progressive Alliance (UPA)–2 Government began work on a new regulatory legislation for India's nuclear energy programme—the Nuclear Safety Regulatory Authority (NSRA) in 2011. The bill was tabled in the Lok Sabha on 7 September 2011 but lapsed in 2014 with the dissolution of the Fifteenth Lok Sabha. If the incumbent government wishes to empower nuclear regulation in the country, it would need to revive the bill and table it in the Parliament once again.

The NSRA Bill 2011 proposes to dissolve the AERB and replace it with a new regulatory body to be called the Nuclear Security Regulatory Authority to regulate safety of nuclear materials, facilities, and activities in India. The NSRA is to be headed by a chairman, and the central government will constitute a CNS which will review the overall nuclear and radiation safety policies. The CNS will comprise of the Prime Minister, cabinet ministers, the cabinet secretary, Chairman of the AEC, and experts nominated by the government. The chairman and members of the NSRA will be recommended by two separate search committees appointed by the CNS.

The NSRA Bill of 2011 did not adequately address the issue of regulatory independence and transparency. There are four issues at hand—lack of autonomy, government control, oversight of strategic facilities, and the Right to Information (RTI). In March 2012, the department-related Parliamentary Standing Committee on Science and Technology, Environment, and Forests submitted its report on the NSRA Bill of 2011.[72] The committee headed by Dr T. Subbarami Reddy observed that the bill impinges on the functional autonomy of the NSRA[73] and that the DAE should seriously re-examine the provisions of the proposed NSRA.[74] The CNS proposed by the new bill becomes a powerful body represented in majority by members of the central government, including the Chairman of the AEC, with the power to review safety policies and appoint the search committees for nominating the chairman and other members of the NSRA. The central government can remove any member of the NSRA without a judicial inquiry (but after being 'heard'). Through these provisions and through the CNS, the NSRA is made subordinate to the central government, risking serious conflict of interest and compromising on the autonomy of the NSRA. The bill allows the central government to exempt from the NSRA's jurisdiction those facilities it considers vital for national defence and security, without listing or elaborating on what kind of facilities these would be and who would regulate the exempted facilities. Further, the NSRA lies outside the provisions of the powerful Right to Information Act which adds to the issue of transparency and autonomy.[75]

The jury is still out on which of the recommendations made by the various auditors will be taken on board, if and when a bill is again introduced in the parliament. The move to strengthen the functional autonomy of India's regulatory authority will require substantial political and bureaucratic capital to be invested in first amending existing legislations like the Atomic Energy Act of 1962, bringing out a fresh and reworked legislation for a regulatory authority with 'de jure' functional and institutional independence with a statutory backing, and then sustaining such a body through necessary financial and organisational support.

Safety-related Incidents in India

Accidents at NPPs and nuclear facilities are matters of serious concern. They are wake-up calls for rectifying loopholes not only in system and equipment designs but also in regulatory practices and safety and security cultures. After all, the nuclear regulation is a commitment to ensure the safety of personnel, public, and the environment. Quite contrary to claims of India's impeccable record in nuclear safety, accidents from low to high significance have occurred over many years in India's nuclear facilities. The AERB keeps a careful record of all these incidents through its events reporting system. It is mandatory for the operators of NPPs in India to 'promptly' inform the AERB all events/ incidents related to the operational safety of the plants and submit event reports. Under the AERB reporting system, nuclear incidents have been categorised as 'events' and 'significant events' based on where the event stands on the INES safety significance ratings (1 to 7),[76] and its impact on operational safety.

Following the serious accidents at the Narora Atomic Power Station–1 (NAPS–1) in 1993[77] and at Kaiga in 1994,[78] the AERB initiated a comprehensive evaluation of the safety status of the DAE's nuclear installations in 1995 and reported the lacunae in DAE's safety management in its document titled 'Safety Issues in DAE Installations'.[79] Notably, when it was found that some of the issues from the DAE evaluations of 1979 and 1987 had not been yet addressed, the AERB exercised its statutory powers to bring about necessary rectification.[80] This was also the time when India had just signed the CNS in 1994, which created greater incentive to address the prevailing lacunae. By the time the next accident at an Indian nuclear facility took place in 2003 at the Kalpakkam Nuclear Reprocessing Facility,[81] the AERB had already renounced regulatory authority over BARC's facilities in 2000. After these accidents, India's NPPs witnessed several safety and regulatory enhancements that have ensured fewer accidents.

The radiation leak in 2010 in Delhi's Mayapuri[82] metal scrap market raised very serious questions, domestically and internationally (particularly from the IAEA) on India's nuclear safety and security, and regulatory practices. While, the AERB has been defended with the explanation that the Delhi University Chemistry Department— from where the Cobalt-60 was disposed—received the material in 1970, much before the AERB was established and, therefore, the apex regulatory body could not have accounted for it,[83] the incident did necessitate action by the regulatory body. The AERB launched campaigns for discovering more legacy materials, computerisation of radiation material inventories, tighter inspections, and public awareness.[84] In March 2011, the AERB launched its Safety Guide on Security of Radioactive Sources in Radiation Facilities.

Between 2013 and 2016, the AERB has reported about 143 significant events at Indian NPPs.[85] According to the AERB, while all events were reviewed, the most important lessons were drawn from these events:[86]

1. Inadvertent release of tritium activity to storm water drain at NAPS in June 2013
2. Inadvertent Radiation exposure of radiation workers at TAPS–3 and TAPS–4 in May 2014
3. Coolant channel leaks in Kakrapar Atomic Power Station (KAPS)– 2 in July 2015 & KAPS–1 in March 2016
4. Leak from primary coolant system at Rajasthan Atomic Power Station (RAPS)–2 in January 2016

As is clear from the preceding discussion, India's nuclear regulatory regime is still evolving. The regulatory principle practiced has so far been that prevention is better than cure. This approach has focused on design-basis safety and operational nuclear safety practices. In light of novel incidents like the Mayapuri incident or Fukushima, there is a need to reorient the approach to 'beyond-design-basis' in order to assess and enhance mitigation of events as much as their prevention. Moreover, the 'defence-in-depth' principle has been adopted to not only avoid accidents but also to mitigate them in case they occur and keep emergency responses ready.[87]

Crisis Learning in Nuclear Safety and Security

Nuclear accidents often function as a catalyst to re-examine nuclear safety systems and cultures and address the lacunae therein. This has been true for nuclear regulatory systems around the world. The Three Mile Island accident in the United States in 1979, the Chernobyl accident in the Soviet Union in 1986, and the Fukushima Daiichi nuclear disaster in Japan in 2011 are widely considered to be serious nuclear accidents. After each of these nuclear accidents, the DAE in India carried out high-level internal evaluations regarding the safety of the country's nuclear projects. Two separate task forces were set up, one in 1979[88] and the other in 1987, after the Three Mile and Chernobyl accidents respectively, and their confidential reports identified the then-existing crucial deficiencies that required urgent attention in India's NPPs and made recommendations for rectification. Crisis learning for India has mainly entailed self-assessment and adoption of corrective measures to the structural and personnel safety features of NPPs. The 1987 Task Force recommended the development of organisations and procedures for on-site and off-site emergencies, and almost thirty-four off-site emergency exercises were conducted at various locations in India between 1987 and 2000.[89]

The Fukushima disaster renewed the debate on nuclear safety in the country. The AERB, which was already drawing a lot of domestic criticism for inadequate safety of its NPPs, focused on measures for enhancing the resilience of NPPs to cope with extreme external events that exceed the design bases and to strengthen the provisions for mitigation of severe accidents. Via a directive (AERB Directive No.1/2013), the AERB called for the constitution of an Event Review Committee (ERC) to review and assess the safety significance of even those events that are caused by a natural disaster.[90] Some of the short-term, medium-term, and long-term enhancements pertaining to nuclear regulation, as presented in India's national report to the CNS are:

Short-term measures
1. Review and revision of emergency operating procedures.
2. Training and mock-up exercises of operating personnel.
Medium-term measures
1. Strengthening provision for monitoring of critical parameters under prolonged loss of power.
Long-term measures
1. Enhancing Severe Accident Management programme.

2. Creation of on-site emergency support centre capable of withstanding extreme flood, cyclone, and earthquake etc.

Looking Ahead: Expansion and Challenges

As pointed out earlier, India's nuclear sector has shown a great deal of potential for growth since the India-US nuclear deal (and other deals thereafter) as well as the NSG exemption. The opening up of the civil nuclear market in India is estimated to have a business potential of USD 60–100 billion.[91] By 2015, nuclear power production in India had already doubled compared to the previous five years and is expected to double further till 2020.[92] Foreign collaborations on nuclear reactors and fuel will play a significant role in providing an upward surge to India's nuclear energy industry.

At least four regulatory challenges will become more pronounced as India goes ahead with its expansion and diversification programme— nuclear waste management, transportation safety and security, managing public perceptions, and nuclear security.

India's waste management practices are currently understood to be in line with the IAEA requirements.[93] Beyond the national guidelines, a long-term nuclear waste management strategy for systematic and coordinated safe disposal of radioactive and nuclear wastes has been recommended by regulatory reviewers including the IRRS. Incidents like the Mayapuri radiation leakage have been wake-up calls for the DAE as well as the AERB and a lesson in improving inter-agency and inter-facilities coordination for radioactive waste disposal. An expansion of the nuclear energy sector in India also implies that greater amounts of nuclear and radioactive materials would be moving from one place to another—from the suppliers to the facilities, from facilities to waste disposal sites, or from facilities and storage sites to safe places in case of emergencies.

The challenge for the regulatory authority consists of formulating adequate guidelines and standards for transportation safety and security of radioactive materials depending upon the fissile characteristic of the material, and ensuring compliance from safety managers through regulatory reviews. The AERB has safety guidelines, manuals, and codes on safe transport of radioactive material, fashioned upon the IAEA regulations.[94]

Any regulator responsible for ensuring the safety of nuclear materials and facilities, has the added responsibility of managing public perception on nuclear energy. Nuclear is an important source of energy, however the disastrous effects of nuclear accidents have created fear and resentment among publics around the world toward nuclear facilities, especially among communities living close to power stations, communities dislocated by nuclear power projects, and those previously affected by radioactive leakages and accidents. Indeed, there is a robust anti-nuclear movement in the country, much of which is against nuclear power plants rather than against nuclear weapons. A great deal of this uneasiness comes from a lack of proper communication strategy and inadequate compensation for the land acquired for building power plants. The regulatory body in India will need to ensure that the nuclear power industry and infrastructure in the country adhere to the rules, regulations, and practices of the highest standards so as to generate public confidence in the envisaged expansion of the nuclear power sector in India.

Finally, although nuclear security is different from safety and safeguards, the three are not mutually exclusive. There are multiple organisations in India involved with the different aspects of nuclear security.[95] As the oversight agency, the regulatory body has to prescribe the technical basis and standard operating procedures for establishing security at various levels—called the graded approach. While the AERB currently prescribes such standards and also reviews the security practices at multiple levels, the absence of an overarching authority that dedicatedly looks after the nuclear security apparatus in India results in a non-uniform security culture.[96]

Notes and References

1. 'Nuclear Power Plants', Atomic Energy Regulatory Board, last modified 19 March 2019, accessed 10 September 2017 <http://www.aerb.gov.in/index.php/english/regulatory-facilities/nuclear-power-plants>

2. 'Growth in Atomic Energy Sector', Lok Sabha, Unstarred Question 343, last modified 25 February 2015, accessed on 10 September 2017 <http://dae.nic.in/writereaddata/parl/budget2015/lsus343.pdf>

3. 'Annual Report 2016–17', Department of Atomic Energy, Government of India, p. 3 <http://dae.nic.in/writereaddata/areport/ar2016_17_eng.pdf>

4. 'India's energy consumption to grow faster than major economies', *Economic Times*, 27 January 2017, accessed on 25 October 2017 <https://economictimes.indiatimes.com/industry/energy/oil-gas/indias-energy-consumption-to-grow-faster-than-major-economies/articleshow/56800587.cms>

5. 'Nuclear Energy and Societal Development', DAE, Government of India, p. 3 <http://dae.nic.in/writereaddata/.pdf_42>

6. 'Meeting Demand Projection', DAE, accessed on 1 November 2017 <http://www.dae.nic.in/?q=node/129>

7. Kanishka Jayasuriya, 'Emergence of Global Regulatory Governance', *Global Legal Studies Journal*, vol. 6, no. 2 (1999), p. 454.

8. International Atomic Energy Agency (IAEA), INSAG 21, 'Strengthening the Global Nuclear Safety Regime', 2006, p. 1, <http://www-pub.iaea.org/MTCD/Publications/PDF/Pub1277_web.pdf>

9. International Atomic Energy Agency, *IAEA Safety Glossary: Terminology in Nuclear Safety and Radiation Protection* (Vienna: IAEA, 2007), p. 164 <https://www-pub.iaea.org/MTCD/publications/PDF/Pub1290_web.pdf>

10. Ibid., p. 4.

11. 'Convention on Nuclear Safety: Introduction to the CNS and Its Associated Rules of Procedure and Guidelines' (IAEA, 2017), p. 43 <https://www-ns.iaea.org/downloads/ni/safety_convention/related-documents/cns-brochure_final_2017-01-23.pdf>

12. 'Legislative and Regulatory Framework', IAEA, NS Tutorials, accessed on 4 October 2017, <https://www.iaea.org/ns/tutorials/regcontrol/legis/legis135.htm>

13. IAEA, 'Nuclear Security Culture', *Nuclear Security Series* 7 (2008), p. 5.

14. Ibid., p. 3; Where, nuclear security is 'the prevention and detection of, and response to, theft, sabotage, unauthorized access, illegal transfer and other malicious acts involving nuclear and other radioactive substances or their associated facilities.' It should be noted that nuclear security includes physical protection as understood in the CPPNM and its Additional Protocol.

15. Ibid., p. 5.

16. R. Baldwin, C. Scott and C. Hood (eds.), *A Reader on Regulation* (Oxford: Oxford University Press, 1993), p. 3.

17. IAEA, *Safety Glossary*, 2007, p. 42.

18. These include fuel fabrication centres, reprocessing plants, nuclear medical centres and even private medical clinics, hospitals, and laboratories using diagnostic and therapeutic equipment containing radioactive sources such as the x-ray machines.

19. 'History', IAEA, accessed on 25 September 2017 <https://www.iaea. org/about/overview/history>

20. Jayasuriya, 'Emergence of Global Regulatory Governance', p. 427.

21. IAEA, INSAG–21, p. 3.

22. Ibid.

23. 'International Conference on Effective Nuclear Regulatory Systems—Facing Safety and Security Issues', IAEA, Conference proceedings, Moscow, 27 February–3 March 2006 <http://www-pub. iaea.org/iaeameetings/50799/ International-Conference-on-Effective-Nuclear-Regulatory-Systems-Sustaining-Improvements-Globally>

24. IAEA, International Conference on effective Nuclear Regulatory Systems—Sustaining Improvements Globally, Conference proceedings, Vienna (Austria), 11–15 April 2016, p. 6 <http://www-pub.iaea. org/MTCD/Publications/PDF/ PUB1790_web.pdf>

25. IAEA, Power Reactor Information System (PRIS), 'The Database on Nuclear Power Reactors', accessed on 14 November 2017 <https://www. iaea.org/pris/>.

26. Ibid.

27. 'World Statistics Nuclear Energy Around the World', Nuclear Energy Institute, accessed on 5 September 2017 <https://www.nei. org/Knowledge-Center/Nuclear-Statistics/World-Statistics>.

28. IAEA, 'Convention on Nuclear Safety: Introduction to the CNS and Its Associated Rules of Procedure and Guidelines', 2017, p. 36 <https://www-ns.iaea.org/ downloads/ni/safety_convention/ related-documents/cns-brochure final_2017–01–23.pdf>.

29. Sitakant Mishra, *Defence beyond Design,* New Delhi, Knowledge World, 2016, p. 18.

30. Ibid., p. 38.

31. Rajiv Gandhi, 'A World Free of Nuclear Weapons: An Action Plan', Speech by the Prime Minister of India, New York, Third Special Session on Disarmament, United Nations General Assembly, 9 June 1988 <http://www.bearve.net/ blog/speeches/rajiv-gandhi-speaks-aagainst-nuclear-weapons>

32. 'Standstill Agreement', *Indian Pugwash Society*, accessed on 2 November 2017 <http://www. pugwashindia.org/Issue_Brief_ Details.aspx?Nid=73>

33. Earlier, India had raised four major objections to the CTBT: 'First, there was no provision for nuclear disarmament in the treaty; second,

the treaty was not comprehensive because it allowed non-explosive tests; third, the operation of the treaty would have given the countries that had already tested an advantage, thus creating a new discriminatory divide; and fourth, the Annex 2 of the treaty where 44 countries, including India were placed. India maintains that Annex 2 of the treaty is against the Vienna Convention on the Law of Treaties', See: Rajiv Nayan, 'The Global Nuclear Non-proliferation Paradigm and India', *Strategic Analysis*, vol. 35, no. 4 (2011), pp. 560–1.

34. 'India links CTBT with disarmament', *Thaindian News*, 30 March 2009 <http://www.thaindian.com/newsportal/uncategorized/india-links-ctbt-with-disarmament_100173358.html>

35. Jaswant Singh, 'Suo Motu Statement by Minister of External Affairs in Parliament on the Non-Proliferation Treaty Review Conference', *Ministry of External Affairs*, Government of India, 9 May 2000 <http://www.indianembassy.org/policy/npt/jsingh_npt_may_9_2000.htm>

36. The Clinton Administration was vociferously against the Indian nuclear programme and had asked New Delhi to 'cap, roll back and eliminate' its nuclear weapon programme.

37. 'US imposes sanctions on India', *CNN.com*, 13 May 1998 <http://edition.cnn.com/WORLD/asiapcf/9805/13/india.us/>

38. Siddharth Varadarajan, 'IAEA Board approves Indian Additional Protocol', *The Hindu*, 5 March 2009 <http://www.hindu.com/2009/03/05/stories/2009030555121100.htm>

39. 'India gets NSG waiver, Manmohan calls it "historic deal"', *The Indian Express*, 6 September 2008 <http://archive.indianexpress.com/news/india-gets-nsg-waiver-manmohan-calls-it-historic-deal/358098/>

40. Siddharth Varadarajan, 'NSG ends India's "clean" waiver', *The Hindu*, 24 June 2011 <http://www.thehindu.com/news/national/nsg-ends-indias-clean-waiver/article2132457.ece>

41. 'Nothing will detract from NSG's clean waiver to India: France', *Two Circles*, 5 July 2011 <http://twocircles.net/2011jul05/nothing_will_detract_nsgs_clean_waiver_india_france.html>; 'India wants NSG partners to "fully honour their commitments"', *The Hindu*, 12 August 2011 <http://www.thehindu.com/news/national/india-wants-nsg-partners-to-fully-honour-their-commitments/article2347435.ece>

42. Mishra, *Defence Beyond Design*, p. 115.

43. Department of Nuclear Safety and Security, 'Integrated Regulatory Review Service (IRRS) Report to India', IAEA, 16–27 March 2015, p. 2 <http://www.aerb.gov.in/images/PDF/25-November–2015.pdf>

44. Ibid.

45. Ibid.

46. Mishra, *Defence Beyond Design*, p. 115.

47. A. R. Sundararajan et al., *Atomic*

Energy Regulatory Board—25 Years of Safety Regulation (Government of India, 2008), p. 23.

48. A. Gopalakrishnan, 'Evolution of the Indian Nuclear Power Program', *Annual Review of Energy and the Environment* 27, November 2002, pp. 384–5.

49. Ibid., p. 378.

50. Ibid., p. 385.

51. 'Policies Governing Nuclear and Radiation Safety', *Legal Pundits*, p. 3, accessed on 13 October 2017 <http://www.legalpundits.com/Content_folder/policiesgrnrs26082014.pdf> (site discontinued).

52. These include uranium and thorium mining and processing (including uranium enrichment), heavy minerals mining and processing, uranium and thorium fuel fabrication, heavy-water production, nuclear power plants and research reactors, spent fuel processing, plutonium recycling and fuel fabrication, zirconium, beryllium extraction and processing, nuclear and radioactive materials waste management, isotope handling and processing.

53. Radioactive sources in radiation facilities in medical, industrial and research institutions including diagnostic medical X-ray installations, radiation therapy installations, nuclear medicine laboratories, industrial radiography installations, gamma irradiation plants, nucleonic gauges, and consumer products.

54. 'Website Glossary', AERB, last modified on 13 August, accessed on 13 October 2017 <http://www.aerb.gov.in/index.php/english/website-glossary>

55. AERB, 'Policies Governing Nuclear and Radiation Safety', *Legal Pundits* <https://aerb.gov.in/PDF/Policies_Governing_Regulation.pdf>

56. Ibid., pp. 1–2.

57. 'Code of Ethics', AERB, last updated on 12 August 2017, accessed on 13 October 2017 <http://www.aerb.gov.in/index.php/english/about-us/code-of-ethics>

58. Government of India, 'National Report to the Convention on Nuclear Safety—Seventh Review Meeting of Contracting Parties', March 2017, pp. 41–42.

59. IAEA, Office of Legal Affairs, 'Country Details—Republic of India', last updated 25 March 2017, accessed on 15 October 2017 <https://ola.iaea.org/ola/FactSheets/CountryDetails.asp?country=IN>

60. Press Information Bureau, '23 Emergency Response Centres of Department of Atomic Energy operational at various sites of DAE', DAE, 6 May 2015 <http://pib.nic.in/newsite/PrintRelease.aspx?relid=121282>

61. DAE, 'Annual Report 2016–17', III.

62. Government of India, 'National Report to the Convention on Nuclear Safety Seventh Review Meeting of Contracting Parties', March 2017, p. 1.

63. DAE, 'Global Centre for Energy

Partnership—Mission', Government of India <http://www.gcnep.gov.in/about/about.html>.

64. 'L&T builds India's first 700 MWe steam generator for nuclear power plant', *Business Standard*, 3 July 2015 <http://www.business-standard.com/content/b2b-manufacturing-industry/l-t-builds-india-s-first–700-mwe-steam-generator-for-nuclear-power-plant–115070300689_1.html>

65. Mishra, *Defence Beyond Design*, p. 36.

66. A. R. Sundararajan, K. S. Parthasarathy and S. Sinha (eds.), *Atomic Energy Regulatory Board—25 Years of Safety Regulation*, Government of India, 2008, p. 23 <http://www.aerb.gov.in/images/PDF/Silver_Jubilee_Book/contents.pdf>

67. Atomic Energy Act, 1962, Government of India, 15 September 1962; AERB, 'Policies Governing Nuclear and Radiation Safety' *Legal Pundits*, p. 1.

68. S. Anand, 'India's Worst Radiation Accident', *Outlook*, 28 July 2003, pp. 18–20.

69. Harsimran Kalra, 'Nuclear Safety Regulatory Authority', *PRS Legislative Brief*, 28 September 2012, p. 2, accessed on 27 September 2017 <http://www.prsindia.org/uploads/media/Nuclear%20Safety/Legislative%20Brief%20NSRA%20Bill,%202011.pdf>

70. Happymon Jacob, 'Regulating India's nuclear estate', *The Hindu*, 29 August 2014 <http://www. thehindu.com/opinion/lead/lead-article-regulating-indias-nuclear-estate/article6360984.ece>

71. 'Parliamentary panel slams AERB on radiation safety policy', *The Economic Times*, 18 October 2013 <https://economictimes.indiatimes.com/industry/energy/power/parliamentary-panel-slams-aerb-on-radiation-safety-policy/articleshow/24348508.cms>

72. <http://164.100.47.5/newcommittee/reports/EnglishCommittees/Committee%20on%20S%20and%20T,%20Env.%20and%20Forests/221.pdf>

73. Karan Malik, 'Standing Committee Report Summary—The Nuclear Safety Regulatory Authority Bill, 2011', *PRS Legislative Research*, 14 March 2012, accessed on 27 September 2017 <http://www.prsindia.org/uploads/media/Nuclear%20Safety/SCR%20summary-Nuclear%20Safety%20Regulatory%20Authority%20Bill,%202011%20.pdf>

74. Jacob, 'Regulating India's Nuclear Estate', *The Hindu*.

75. Ibid.

76. Events rated at levels 1, 2, and 3 are called 'Incidents'. Events with no safety significance are rated at level 0 or below scale.

77. Fire incident at NAPS–1 in March 1993.

78. Fall of the construction dome at Kaiga in May 1994.

79. A. Gopalakrishan, 'Evolution of the Indian Nuclear Power Program', p. 386.

80. Ibid.

81. Radioactive leakage.

82. In April 2010, it was reported that a scrap dealer in Delhi's Mayapuri industrial area died and five others suffered burn injuries due to accidental exposure to radiation. Subsequent enquiries found that the victims were exposed to the highly radioactive and powerful isotope—Cobalt 60—used normally for medical and industrial purposes, that was left in the market as a scrap consignment.

83. Mishra, *Defence Beyond Design*, pp. 133–4.

84. AERB, 'Annual Report 2010–11', Government of India, p. 4.

85. AERB, 'Annual Report 2013–14', Government of India; AERB, 'Annual Report 2015–16', Government of India; and 'Significant Events', AERB, accessed on 14 October 2017 <http://www.aerb.gov.in/images/PDF/npp/significantevents.pdf>

86. Government of India, 'National Report to the Convention on Nuclear Safety Seventh Review Meeting of Contracting Parties', March 2017, pp. 19–21.

87. Baldev Raj, 'Nuclear Energy After Major Accidents: Economics, Safety, Risk, Scale, Acceptance', Lecture on 'Global Nuclear Politics and Strategy', Bangalore, Vth Annual Residential Workshop for Young Scholars, International Strategic and Security Studies Programme (ISSSP), NIAS, 7 May 2015 <http://isssp.in/wp-content/uploads/2015/06/13-Baldev-Raj-Nuclear-Energy-After-Major-Accidents.pdf>

88. In June 1979, the DAE's Secretary, Homi. N. Sethna constituted a taskforce under the Chairmanship of M. R. Rao, the then Head, Reactor Operation Division at BARC, to study in detail, the safety aspects of TAPS and RAPS and come up with early recommendations. The report submitted by the taskforce was discussed by SRC in an extended meeting, held during October 1979.

89. 'Nuclear Emergency Response', DAE, accessed on 14 October 2017 <http://dae.nic.in/?q=node/37>

90. AERB, Directive No.1/2013, 'Criteria and Assessment Procedures for Notification of Nuclear Incident Under the Civil Liability for Nuclear Damage Act, 2010', 9 December 2013 <http://www.aerb.gov.in/images/PDF/directive2013civilnuclearliability.pdf>

91. Anil Sasi, 'Nuclear Safety Regulatory Authority Bill: Statutory backing key to better safety', *The Indian Express*, 26 April 2017 <http://indianexpress.com/article/business/business-others/nuclear-safety-regulatory-authority-bill-statutory-backing-key-to-better-safety-4628342/>.

92. Amit Bhandari, 'Why India's nuclear power output is surging,' *Business Standard*, 3 February 2015 <http://www.business-standard.com/article/economy-policy/why-india-s-nuclear-power-output-is-surging-115020300354_1.html>

93. Aniruddha Mohan, 'Nuclear Safety

and Regulation in India: The Way Forward', *ORF Issue Brief* 96, June 2015, accessed on 27 September 2017, p. 6 <http://www.orfonline.org/research/nuclear-safety-and-regulation-in-india-the-way-forward/>

94. 'Transport of Radioactive Material', AERB, last updated on 15 August 2017, accessed on 26 October 2017 <http://www.aerb.gov.in/index.php/english/regulatory-facilities/transport-of-ram>

95. Happymon Jacob and Sitakanta Mishra, 'Nuclear Security Governance in India: Institutions, Instruments, and Culture,' Sandia Report, SAND2015–0233, *Sandia National Laboratories*, p. 54, accessed on 23 August 2017 <http://www.sandia.gov/cooperative-monitoring-center/_assets/documents/sand2015–0233.pdf>

96. Ibid.

6

India's Ballistic Missile Defence Programme: Implications for Strategic Stability

Naeem Salik

Brief Historical Background

The history of efforts to develop an antidote to ballistic missiles is almost as old as the advent of modern ballistic missiles themselves. Towards the closing stages of the Second World War, the Germans employed two types of missiles, the V-1 and V-2, against London and other targets in Britain. While the V-1 was a cruise missile with a powered flight, the V-2 was a ballistic rocket. These weapons, though not very accurate and sophisticated, were the forerunners of the modern day ballistic missiles of varying ranges from short range missiles to those with intercontinental reach. During the 1950s, President Eisenhower ordered the initiation of the development of a nuclear-tipped interceptor missile named 'Nike Zeus'. Later on, the scope of the research was expanded to the development of a nation-wide missile defence system. Concurrently, the Soviets were pursuing their own ballistic missile defence (BMD) system which led to their deployment around Moscow. President Nixon responded by authorising the deployment of the 'Safeguard' anti-ballistic missile system in the late 1960s despite serious doubts about its efficacy.

The technological deficiencies of BMDs make these weapon systems highly destabilising. Since every BMD system has a limited capacity to intercept incoming missiles and can be overwhelmed by a full blown strike, the side having deployed a BMD system would be tempted to go for a first strike to reduce the size of the hostile missile strike to enable its defence system to handle it effectively. The side without BMDs would be under pressure to launch its missiles before its capability is degraded by a pre-emptive strike by the adversary. This will create what is called the first strike instability in any crisis situation. Recognising the destabilising potential of BMD systems, the two super powers started negotiations to limit the deployments of such defensive systems as part of the Strategic Arms Limitation Talks (SALT) negotiations process. As a result, the Anti-Ballistic Missile (ABM) Treaty was signed in 1972, which initially allowed two sites with 100 interceptors each to either party. This was further curtailed to 100 interceptors at one site each, which could either be the capital city or an intercontinental ballistic missile (ICBM) site.

President Reagan's March 1983 Strategic Defence Initiative (SDI) speech gave a new fillip to the research and development of BMD systems with billions of dollars budgeted for it. However, with the end of the Cold War, the scope of the development was curtailed with reduced budgetary allocations. In May 2001, President George W. Bush announced his plan to deploy the National Missile Defence (NMD) and in June 2002, he announced unilateral US withdrawal from the ABM Treaty that had been considered one of the critical elements for maintaining strategic stability between the US and erstwhile Soviet Union and had facilitated the SALT process.

India had been critical of Reagan's SDI programme and considered it 'disruptive';[1] however, when President Bush announced the NMD plan and later abrogated the ABM Treaty, it proved to be a more enthusiastic supporter of Bush's plans than long-standing US allies such as Japan and even its NATO partners. India's change of heart occurred after a tele-conversation between the US National Security Advisor Condoleezza Rice and Indian Minister of External Affairs Jaswant Singh.[2] Within India, the leftist political parties were highly critical of this action of the BJP government. In an article in the weekly publication of the Communist Party of India titled the 'People's Democracy', Prakash Karat wrote that, 'The speed with which the Vajpayee government uncritically welcomed President Bush's new missile defence system and changes in strategic nuclear policy have taken aback even those who broadly support the pro-US orientation in foreign policy...'.[3] In his concluding comments, he added that:

> The Bush administration has embarked on a new aggressive global strategy which is causing unease even among its allies. The Vajpayee government has put up a display of unashamed sycophancy in its anxiety to worm its way into the good books of America. In this process, it has announced to the world what has been evident for some time now. India, under BJP rule, aspires to the status of a regional cog in the American global strategy.[4]

India's Quest for BMD—Political and Strategic Drivers

India saw in Bush's scheme a reordering of the old international order wherein it had perceived itself being denied its rightful place and saw an opportunity to carve a more advantageous position in the new order by aligning itself with the policies of the predominant global power. Urging the Indian leadership to grab this opportunity, a prominent Indian security analyst, C. Raja Mohan, eloquently argued that:

> There is hardly any reason why India should be shedding tears at the demise of the old nuclear order. President Bush's plan, which opens the door for a rewriting of the rules of the nuclear game, offers India a chance to be part of the nuclear solution and not the proliferation problem. It makes all sense for India to lend a helping hand to the United States to dismantle the current nuclear regime, which New Delhi has long called discriminatory.[5]

This transition was also facilitated by the common goal of both the US and India to manage the rise of a resurgent China. India also saw in missile defence cooperation an opening of the door to cutting-edge US military technologies that it had yearned for long. India has also been seeking status through its technological advancements, especially in the nuclear, missile, and space domains and strategic experts such as K. Subrahmanyam had long been terming nuclear weapons as currency of international power. Taking a similar line, another Indian expert advocating India's embrace of the BMD technology has stated that:

> ICBMs are already an idiom of big power status ... power projection assets also include a hardy naval force of indigenously built submarines, instruments of precision strikes in the form of cruise missiles, and space based information systems. Missile defences will put India on that path.[6]

Charles Ferguson and Bruce Macdonald have also pointed out that in addition to the perceived missile threats, 'prestige is an additional rationale for BMD development.... Development of BMD, a complex military technology, would show that India is a member of an elite club.'[7]

Brahma Chellaney, justifying the need for a missile defence shield for India has, in a rather paranoid threat assessment, counted China, missile armed foreign naval fleets in the Indian Ocean, Pakistan, Iran, and even old vintage Chinese supplied intermediate-range ballistic missiles (IRBMs) of Saudi Arabia amongst the potential missile threats to India.[8] He also suggested that it would be prudent for India to support the NMD programme of the United States as a means of further promoting its friendship with the US adding that, 'To see the Chinese rattled by US missile defences is surely an agreeable sight for India'.[9] He, however, acknowledged the fact that likely Chinese responses to US deployment of an NMD system will have adverse effects on Indian security.[10] But he saw in it the establishment of a new

US-led security framework which could include a 'missile umbrella' similar to the nuclear umbrella extended by the US to its allies and India could also benefit from such an arrangement.[11]

The timeline of the initiation of India's BMD quest is not very clear. However, most studies trace these from the mid- to late-1990s.[12] Charles Ferguson and Bruce MacDonald contend that the programme officially commenced with the pronouncement of the 'Defence for Delhi' policy, which included defence of Mumbai as well, 'due to the significant number of nuclear facilities near there, though BMD research had started some years earlier. They also believe that nationwide missile defence is beyond India's capacity.'[13] A major turning point in India's policy towards BMD came as a result of its public support of President Bush's NMD plan. India saw this alignment with US policy as a means of gaining access to advanced technologies including a possible transfer to India of BMD technology by the US. It also saw in it an opportunity to be recognised as a legitimate player in the new global nuclear regime that was likely to emerge once Bush's efforts to dismantle the existing order reached fruition. Some Indian analyst argued that India did not have the means to counter a nuclear attack from either of its two adversaries—Pakistan and China—and therefore it made sense for India to become part of the larger US missile defence architecture.[14]

Ambassador Robert Blackwill, the US ambassador to New Delhi, invested a lot of his political clout in Washington to broaden and deepen the US–India cooperation that now included missile defence as a part of the four major areas of collaboration as reflected in the Next Steps in Strategic Cooperation (NSSP) announced in January 2004. Till that time, the Defence Research and Development Organisation's activities had been confined to conceptualisation of India's ballistic missile defence requirements and exploratory talks with countries like Russia and Israel for the acquisition of air defence systems, airborne early warning aircraft, and missile launch warning and tracking radars.[15]

Tellis has argued that India's decision to embrace BMD was influenced by the 9/11 incident and the fear that as a consequence of the US military operation Enduring Freedom in neighbouring Afghanistan, Pakistan's security would be degraded, including the possibility of a loss of control over its nuclear arsenal.[16] Another strategic analyst of Indian origin, Sumit Ganguly, has in a somewhat similar vein pointed out that India's prospective BMD system is 'designed to protect India's national capital area and its command and control facilities from a "rogue" nuclear strike from Pakistan ...'.[17] Both these arguments are hard to comprehend. Firstly, the 9/11 attacks involved airliners not missiles and does not explain as to how a BMD system could have prevented such an attack. Secondly, it appears that both Tellis and Ganguly seem to imagine that in the unlikely possibility of a loss of control over Pakistan's nuclear arsenal, the 'rogue' elements would be able to not only sneak away some warheads but ballistic missile delivery systems as well and that would require India to have operational BMD system to defend itself against such an unauthorised strike. It somehow appears to be a scaled down model of US logic for deploying NMD against a strike supposedly by countries like Iran and the Democratic People's Republic of Korea (DPRK), whom the US has labelled as 'rogue states'. It appears that both scholars have gone out of their way to justify India's development, acquisition, and deployment of a BMD system.

Ganguly has proffered some interesting ideas regarding the main drivers of India's BMD effort. First, in his view is the organisational imperative, which means that the DRDO's desire to stake a claim over a major share of military R&D funds and stay ahead of other competing organisations by embarking upon a highly complex and sophisticate project such as BMD. Second, he believes that the most compelling reason for India's pursuit of BMD lies in, 'a peculiar strategic dilemma confronted by the Indian state, i.e. Irregular Warfare and Nuclear Weapons' being employed by Pakistan against India. A corollary of the second driver, the third driver, is the failure of the

Indian state to come to terms with the challenge posed by Pakistan through the employment of military means. As a result, the 'Indian policy makers are now seeking a technological solution…. This quest for a technological answer has led India to pursue BMD…'.[18] Again, this argument defies all logic and does not stand on firm ground because it is hard to imagine a BMD system being able to deter or counter a 'proxy' war. Beyond this assertion, Ganguly has not elucidated as to how India's acquisition and operational deployment of a costly and sophisticated missile defence system dissuade the non-state actors from their inimical actions against the Indian state. One would have to really stretch the imagination to believe that non-state actors perpetrating small-scale incidents or even more serious actions such as the attack on the Indian Parliament or the Mumbai incident could be deterred by either a limited or an extensive deployment of BMDs by India. However, he might be suggesting that in case India responds to any action on its soil or in Jammu and Kashmir by some reprisal action on Pakistani territory, Pakistan's retaliatory options may be constrained by an operational BMD system in India. But even in case of limited tit-for-tat actions by the two sides, it is hard to imagine either side contemplating the use of ballistic missiles irrespective of the kind of warheads they carry unless they get entangled in a full scale conventional war. So how does this technical solution resolve India's sub-conventional dilemma is hard to fathom.

Tellis on the other hand has identified some of India's interests in joining the American bandwagon on BMD as, 'acceptance in the new global nuclear regime, achievement of equilibrium in US–India–Russian relations and enhanced technology cooperation between India and the United States.'[19] Ganguly, however, proceeds to reveal the real strategic purpose behind India's pursuit of BMDs, hinting that in addition to their role in limiting the damage from a Pakistani nuclear strike, India's BMD capability could be used to support an offensive first strike by India stating that:

Assuming that India did achieve the requisite technical capabilities and had a sufficiently risk-acceptant leadership, the scenario of an Indian first strike may not seem chimerical. After such a strike, which would disable much of Pakistan's nuclear arsenal, the ragged retaliation that would follow could be significantly denuded through the use of India's BMD.[20]

Given India's declared NFU policy, this idea of a pre-emptive first strike may seem sinister but was later echoed by the former Indian Secretary of External Affairs and National Security Advisor, Shiv Shankar Menon, in his book, *Choices* (2016), and then by Defence Minister Manohar Parrikar. The idea has since been further expounded by Vipin Narang, a US-based academic of Indian origin, and appears to be well entrenched in Indian strategic thinking though the declaratory doctrine has not yet been formally modified to reflect this change. In all probability, the declaratory doctrine will remain the same to serve India's politico-diplomatic interests and to maintain the image of a responsible nuclear power.

Recognising the intricacy of the technologies related to various components of a BMD system, India decided to follow a dual track of indigenous research and development in some areas and acquisition of technologies from abroad in the other, though the major domestic stakeholder (DRDO) was opposed to the import of BMD systems and sought a bigger share of funding for BMD research and development.[21] It also entered into collaboration with Israel and France in the development of target tracking and the fire control radar 'Swordfish'. India has also acquired Green Pine Radar, part of the Arrow missile defence system from Israel.[22]

Indo–US Cooperation in BMD

The formal Indo–US collaboration in BMD commenced in May 2002 during the Defence Policy Group meeting in New Delhi wherein the Indian Defence Secretary Yogendra Narain discussed the issue in some detail with his American counterpart Douglas Feith and hinted at the Indian political leadership's readiness to accept missile defence as a panacea for its peculiar strategic problems. The two sides also discussed the modalities and areas of US technical assistance. The US side offered to hold a workshop in New Delhi in 2003 to help the Indian side assess its missile defence requirements. The Indian side agreed to take part in a BMD-related table top exercise at Colorado Springs towards the end of May 2002, accepted the offer to attend a missile defence conference at Dallas scheduled for June 2002, and agreed to witness the 'Roving Sands' missile defence exercise at New Mexico in June 2003.[23] During a visit to New Delhi by Deputy Assistant Secretary of Defence David Trachtenberg in January 2003, his Indian interlocutor, Sheel Kant Sharma, apprised him of the Indian government's decision to incorporate BMD into India's defence architecture.[24] When President George W. Bush announced the NSSP in January 2004, he pronounced the agreement between the US and India to broaden the scope of their consultations on missile defence which opened the possibility of a prospective transfer of BMD systems to India.[25]

India's Efforts to Develop an Indigenous BMD System

According to Sumit Ganguly, India is trying to build a twin-layered missile defence system that includes a high altitude interceptor, Prithvi Air Defence Missile (PAD), and a low altitude interceptor, Ashwin Advanced Air Defence System (AAD).[26] American experts have argued that India is seeking a 'two-tiered' missile defence due to the nature of a missile threat from Pakistan and China and their respective geographical configurations, which means that Pakistani missiles pose short to medium range threats while the Chinese missile threats fall in the intermediate range category.[27] India started its testing of indigenous missile defence interceptors in November

2006, by conducting the first test of the PAD missile and that of the AAD missile in December 2007.[28] The series of tests continued with another test of PAD in March 2009 using a ship-launched Dhanush missile as the target missile. This test was supported by the 'Swordfish' Long Range Tracking Radar.

In April 2014, the DRDO tested the Prithvi Defence Vehicle (PDV)—the successor to the PAD missile. The PDV is designed to intercept the incoming missiles in the exo-atmospheric range. This test was reportedly a 'near miss'; however, the DRDO declared it as a successful intercept. In February 2017, a 2,000-kilometre range missile fired from an Indian ship in the Bay of Bengal was intercepted and in early March 2017, an endo-atmospheric interceptor engaged an incoming missile at an altitude ranging between 15–25 km. This test was also declared as successful. The DRDO has claimed the capability of intercepting missiles with ranges up to 2,000 km and announced its plans to engage missiles with ranges of 5,000 km.[29] However, the DRDO's claim should be seen in the context of its past record of making exaggerated claims about its achievements.

In the light of the ratio of successful intercepts claimed by the Missile Defence Agency of the United States with access to far more sophisticated technologies as compared to the DRDO, the successful intercept rates claimed by the DRDO are incredible to say the least. After conducting only three tests, the DRDO boldly claimed that its programme 'has reached a sufficient maturity level to engage'. The recent agreement signed by the Indian government with Russia for a multi-billion dollar purchase of the S-400 Air Defence System is a clear acknowledgement that the DRDO's efforts, spanning over two decades to develop an effective BMD system, have not been successful—contrary to its exaggerated claims of success. The bulk of India's existing air defence assets are obsolescent and need a major refurbishing/ replacement. The induction of the S-400 would solve many of its air defence challenges. However, it will create another headache. The synchronisation of old vintage air defence assets, indigenous BMD systems, and the S-400 into one cohesive defence architecture would pose very serious technological and management problems.

Counter-measures against Ballistic Missile Defence Systems

A whole menu of counter measures is available against BMD systems. The simplest of these being the multiplication of offensive missile numbers to overwhelm the defensive systems. Other counter measures are built into the payloads of ballistic missiles. These include decoys, metallic chaff, and other penetration aids. Then there are more advanced technological solutions such as multiple independently targetable re-entry vehicles (MIRVs) and manoeuvrable re-entry vehicles (MaRVs).[30] The other options available are in the form of variations of missile launch techniques such as use of 'lofted trajectories' by land-launched missiles or 'depressed trajectories' followed by submarine launched ballistic missiles (SLBMs). Then is the option of increasing reliance on cruise missiles which, due to their terrain-hugging flight pattern, are hard to detect by the target identification and tracking radars of the BMD systems.[31] The current variants of cruise missiles also employ stealth technologies, making it even more difficult to detect these. Though the Indian media has projected that the S-400 can also intercept cruise missiles because its 3D acquisition radar can track, among other targets, cruise missiles as well,[32] this claim needs to be viewed with some scepticism.

Moreover, the S-400's longest range interceptor missile, 40N6, has a slant range of 400 km but its maximum altitude is just 30 km, which means that it is incapable of intercepting medium- or long-range missiles which travel most of their flight path in space far above the ceiling of 40N6. Therefore, the S-400 defence system can at best intercept short-range tactical ballistic missiles as has also been shown in its possible targets that include fighter aircraft, smart bombs, cruise missiles, drones, and tactical ballistic missiles.[33] It is clear that this system, which is basically an air defence system designed to shoot down adversary aircraft at long ranges, would not be effective against most Pakistani ballistic missiles that fall in the category of medium-range missiles with apogees much higher than the upper ceiling of the S-400's long range interceptor missile. It appears that the euphoria in Indian media and among Indian experts is based on highly exaggerated claims of the system's capabilities, which is a dangerous trend. Dr Yousuf Butt, an American missile and space technology expert has cautioned against the dangers of overstatement of the BMD systems' capabilities stating that, 'Exaggerating the abilities of missile defence is downright dangerous and military leaders ought to make sure that it doesn't happen; unfortunately it does.'[34] He further elaborates his point, arguing that:

> Unfounded claims of missile defence's effectiveness create a serious risk that the political leaders might be misled into mistakenly believing that missile defences actually work. And if they incorrectly think that missile defence has secured the country by neutralising the threat of ballistic missile attack, policy makers might be emboldened to stake out riskier and more aggressive regional policies than in the absence of missile defence.[35]

Butt also argues that while a tactical missile defence system, such as the US Patriot system, designed to intercept short-range missiles armed with conventional warheads even with a 70 per cent interception capability makes sense, the same cannot be said of the NMD aimed at intercepting long-range nuclear armed missiles. Successfully intercepting seven out of ten nuclear-tipped missiles and allowing the remaining three to hit their intended targets with each one of these remaining warheads capable of causing widespread destruction does not provide much relief. Strategic missile defence 'would need to intercept 100 per cent of incoming nuclear warheads—an unattainable goal for any piece of machinery.'[36]

He considers the BMD system analogous to 'hurricane insurance' explaining that, 'people endanger their lives and property on a regular basis by building on unsafe ground in the knowledge that they are covered for catastrophic events.'[37] He also contends that, 'there is always a reasonable probability that one or more nuclear missiles will penetrate even the best missile defence system.'[38] Other experts like Nobel Laureate Thomas Schelling, commenting on the viability of a prospective US missile defence shield, has said that, 'missile defence will be of dubious value in addressing the possible future threats from Iran.'[39]

Destabilising Impact of Indian BMD Systems and Pakistan's Response

India's deployment of BMD systems would adversely affect regional strategic stability in South Asia in many different ways. First of all, it would lead to arms race instability wherein Pakistan, feeling apprehensive of erosion of the credibility of its nuclear deterrence, could simply build more nuclear warheads and ballistic missiles to ensure that it retains its ability to cause unacceptable damage in case of a deterrence breakdown. It could also start producing more nuclear capable cruise missiles. Besides these quantitative measures, there would also be qualitative solutions such as addition of penetration aids, decoys, etc. in the missiles. All these would trigger an undesirable arms race in the region. Secondly, this Indian move would result in crisis instability because having deployed BMD systems, there will be a temptation on India's part to launch a pre-emptive first strike to degrade Pakistan's offensive capability in the belief that its missile defences would be able to cope with the residual Pakistani capability. There is already a vibrant debate raging in India where current and former senior officials have been advocating scrapping the NFU policy in favour of a first strike nuclear policy. This would create undue pressure on Pakistan to use its assets before these are damaged by an

Indian first strike, leading to a first strike instability or crisis instability. The crisis instability would be further accentuated because the Indian leadership would be tempted to be more prone to brinkmanship and risk taking in a crisis situation, thinking that they are sitting safely behind their missile defence shield.

Signs of this bravado are already visible in the public debate on Indian media and in the statements by the Indian leadership in the immediate aftermath of the inking of the Indo–Russian agreement for the acquisition of S-400 systems despite the fact that the weapon systems would be delivered after a few years. It will also be some time before these are operationally deployed and integrated with the existing Indian air defence architecture. Similarly shrill pronouncements were heard after the inaugural 'deterrence patrol' by the first Indian nuclear-powered, nuclear-armed submarine, *Arihant*, indicating an increasingly aggressive posturing by the Indian leadership. The political leadership's belligerence often coincides with the election cycle in India. Unfortunately, such habits, once adopted, are hard to part with later since these transform into instincts with the passage of time.

Many experts concede that even before the deployment of Indian BMDs, the adverse impact on strategic balance between India and Pakistan is already manifesting itself and, though they believe that it will not have any meaningful impact on Chinese strategic capability,[40] some Chinese counter-measures cannot be ruled out. Pakistan was being blamed by the international media, academia, and think tanks—especially in the US—for rapidly expanding its nuclear arsenal and the 'fastest growing nuclear arsenal' had become a well worn-out cliché. However, this is not a mindless pursuit of a large nuclear arsenal and the increase in Pakistani assets had a basis in strategic logic and security compulsions. According to Rajeshwari Pillai:

> Pakistan is clearly worried about the possibility that India may build or buy an effective BMD shield .... Pakistan has taken a couple of different steps to deal with the problem. One, it is dramatically

increasing the size of its nuclear arsenal .... Second, Pakistan is also developing an entire array of missiles of different ranges to make India's missile defence shield ineffective .... Third, Pakistan is also developing nuclear capable cruise missiles, which cannot be defeated by Indian BMD systems. In brief, India's BMD system may partly be responsible for the acceleration of Pakistan's nuclear arsenal.[41]

Charles Ferguson also concurs with Pillai's assessment, saying that, 'Pakistan's development of nuclear armed cruise missiles appears to be driven in part by Islamabad's desire to counter India's eventual BMD system...'.[42] Another Indian security analyst, Rajesh Rajagopalan of Jawaharlal Nehru University, New Delhi, in an interview with *Arms Control Today*, noted that, 'India does not have the surveillance capacities needed to monitor Pakistan's nuclear forces after they are dispersed, as they presumably would be in a crisis.'[43] This assertion by Rajagopalan brings to light the fact that India's intelligence, surveillance, and reconnaissance (ISR) capabilities are not up to the mark as yet and the temptation to carry out a pre-emptive first strike to supplement its missile defences cannot be fulfilled at present and in the near future.

Pakistan on its part has made it known that it will take all necessary steps to ensure the continued credibility of its nuclear deterrence. It has also signalled that it already has the technological responses to India's acquisition of BMD systems as well as its operationalisation of its first SSBN. Speaking at a conference organised by a local think tank at Islamabad, long serving Director General of Strategic Plans Division (SPD) and currently an advisor to the NCA, Lieutenant General Khalid Kidwai, commenting on India's induction of the S-400 and operationalisation of its SSBN Arihant declared that, 'Pakistan remains unfazed and as before, we have adequate response options which will disallow any disturbance of the strategic balance or strategic stability. That fundamental policy will prevail.'[44]

In a clear reference to the exuberance with which the acquisition of the S-400 system from Russia is being advertised in India, he dispelled the impression being created that the introduction of this system would transform the strategic environment in South Asia stating that, 'much hype has been created around this particular technology induction and some have gone to the extent of calling it a game changer for South Asia,' and went on to dispel this as a wrong impression.[45]

General Kidwai also pointed out that historically, Pakistan has never allowed the strategic balance in South Asia to be disturbed to its detriment and that, 'we have always found effective solutions to redress induced imbalances from time to time.'[46] Elaborating the counter-measures Pakistan has already developed against the Indian BMDs, he stated that among these 'cost-effective solutions' is the MIRV capability as well as four different types of cruise missiles. He also hinted that India's BMD systems are not yet fully operational while Pakistan's response options are already available. Further adding that Pakistan had taken a well-considered decision not to develop its own BMDs.[47] General Kidwai also made it clear that Pakistan's full spectrum deterrence includes a whole range of options at the strategic, operational, and tactical levels which are adequate to deal effectively with INS Arihant.[48]

The message was reiterated by General Zubair Hayat, Chairman Joint Chiefs of Staff Committee who also holds a key position in Pakistan's nuclear hierarchy, while addressing a public event in Islamabad a few days later. Referring to the recent Indo-Russian agreement for the purchase of the S-400 air defence system, he averred that, 'Several contemporary trends and developments pose serious threats to global and South Asian strategic stability. Development and acquisition of missile defence capabilities, nuclearisation of the Indian Ocean ... shall place a great strain on strategic and deterrence stability.'[49] However, he stressed that these developments have neither perturbed Pakistan nor are these unexpected as far as Pakistan is concerned, adding that, 'We will continue to provide necessary response to ensure that strategic balance is maintained and deterrence remains credible.'[50]

Conclusion

India's indigenous development and foreign acquisition of the missile defence systems is a typical case of 'pull of technology' rather than 'push of political policy'. In the case of the United States, the ABM systems development was propped up by the bureaucratic or domestic structure model of the arms race, wherein various military services were competing with each other to retain their role in possessing and operating strategic weapon systems. In India's case, it was not domestic politics but considerations of international politics wherein India was seeking to win US favours by unequivocally supporting its NMD programme. It also saw in this band-wagoning with the US possibilities of gaining access to advanced military technologies. The decision to purchase the costly S–400 system from Russia also had underlying geopolitical considerations. India wanted to maintain a balance in its strategic ties with the United States on the one hand and Russia on the other. It also wanted to keep its traditional arms supplier interested and engaged by signing on to this lucrative deal. To some extent, the bureaucratic politics among the defence research and development organisations vying for greater share in funding also led the DRDO to stake a claim in this area. It succeeded in securing the desired finances and pursued an ambitious programme, going as far as actively opposing any imports of BMD technologies from abroad.

However, in its eagerness to deepen and widen its relations with the United States and open the doors to the acquisition of cutting-edge US technologies besides seeking prestige as a member of the exclusive club possessing BMD technologies, India seems to have been completely oblivious to the destabilising effects of its actions in its relations with its immediate neighbours. In the absence of any overarching restraint mechanism, it has wittingly or unwittingly triggered a strategic arms race in South Asia. Had India carefully studied and absorbed the lessons of the super powers' competition during the Cold War wherein both sides, recognising the destabilising potential of the BMD systems, had signed the ABM Treaty in 1972.

The international security environment post-ABM Treaty also contains lessons for the avid observer, indicating clearly the unsettling doctrinal as well as technological developments in Russia, US, and other major powers. India, at the moment, is happy to play in the big league but it should pause to think that it cannot escape the realities of its immediate security landscape and cannot afford to vitiate it mindlessly.

Notes and References

1. Brahma Chellaney, 'New Delhi's Dilemma', in Alexander T. J. Lennon, ed. *Contemporary Nuclear Debates* (Cambridge, Massachusetts: MIT Press, 2002), p. 122.

2. Ashley J. Tellis, 'The Evolution of US-India Ties—Missile Defence in and Emerging Strategic Relationship', *International Security*, vol. 30, no. 4 (Spring 2006), p. 115.

3. Prakash Karat, 'Accomplice to US Adventurism', *People's Democracy*, vol. 25, no. 19, 13 May 2001.

4. Ibid.

5. C. Raja Mohan quoted in Tellis, op. cit., p. 132.

6. Brahma Chellaney, op. cit., p. 126.

7. Charles D. Ferguson and Bruce W. MacDonald, *Nuclear Dynamics—In a Multipolar Strategic Ballistic Missile Defence World* (Washington, D.C.: Federation of American Scientists, July 2017), p. 13.

8. Ibid., p. 123.

9. Ibid., p. 126.

10. Ibid., p. 125.

11. Ibid., p. 129.

12. For example, Rajeshwari Pillai gives the commencement date in the mid-1990s while Ashley Tellis talks of the late 1990s. See Dr Rajeshwari Pillai Rajagopalan, 'Strategic Implications of India's Ballistic Missile Defence', paper written for FAS Project on 'Nuclear Dynamics in a Multipolar Strategic BMD World', 8 May 2017; Ashley Tellis, op. cit., p. 138; Ferguson and MacDonald, op. cit., p. 11.

13. Ferguson, op. cit., p. 11.

14. Tellis, op. cit., p. 133.

15. Ibid., pp. 137–8.

16. Ibid., pp. 138–9.

17. Sumit Ganguly, 'India's Pursuit of Ballistic Missile Defence', *The Nonproliferation Review*, vol. 21, Issue 3–4 (2014), p. 379.

18. Ibid., pp. 374–7.

19. Tellis, op. cit., p. 136.

20. Ibid., p. 378.

21. Rajeshwari Pillai, op. cit., p. 2.

22. Ganguly, op. cit., p. 373; Ferguson and MacDonald, op. cit., p. 13.

23. Tellis, op. cit., p. 143.

24. Ibid., p. 144.

25. Ibid., p. 146.

26. Ganguly, op. cit., p. 373.

27. Ferguson, op. cit., p. 12.

28. Pillai, op. cit., p. 3.
29. Ibid., p. 4.
30. Ferguson, op. cit., p. 20.
31. Ibid., p. 21.
32. Sundip Unnithan, 'The Geopolitical Missile', *India Today*, 13 August 2018, p. 43.
33. Ibid.
34. Yousuf Butt, 'The Myth of Missile Defence as a Deterrent', *Bulletin of Atomic Scientists*, 8 May 2010.
35. Ibid.
36. Yousuf Butt, 'What Missile Defence?' *Foreign Policy*, 21 October 2009 <http://foreignpolicy.com/2009/10/21/what-missile-defense/>
37. Ibid.
38. Butt, *Bulletin of Atomic Scientist*, op. cit.
39. Schelling quoted in Yousuf Butt, 'The Myth of Missile Defence as a Deterrent', op. cit.
40. Pillai, op. cit., p. 5.
41. Ibid.
42. Ferguson, op. cit., p. 12.
43. Ibid.
44. General Khalid Kidwai, 'Pakistan to Maintain Strategic Balance with India, says NCA Adviser', *Dawn*, Islamabad, 7 November 2018 <https://www.dawn.com/news/1444087>
45. Ibid.
46. Ibid.
47. Ibid.
48. Ibid.
49. 'Pakistan to Ensure Strategic Balance: CJCSC', *The Nation*, Islamabad, 10 November 2018.
50. Ibid.

Conclusion

Naeem Salik

Like any other nuclear weapon state, India has learnt to manage its nuclear capability through its habituation with the 'bomb' over the past two decades. However, much of this learning falls in the domain of 'experiential learning', which is a kind of learning by doing. In the aftermath of nuclear tests and declaration of its nuclear status, India—unlike Pakistan—did not have to start from scratch since it already had a body of indigenous literature accumulated over the years, mainly due to the efforts of strategic experts such as K. Subrahmanyam,[1] military men like General K. Sundarji,[2] and aviator-turned-scholar, Air Commodore Jasjit Singh.[3] There were some other academics as well as military professionals, notable among them was Brigadier V. K. Nair[4] who had been exploring key aspects of nuclear policy such as threat assessment, strategic options, command and control, and targeting options in addition to calculating the damage resulting from a nuclear exchange, and economic implications of nuclearisation. Sundarji also explored various scenarios of a future war in a nuclear environment through table-top exercises. Subrahmanyam later played a key role as head of the first NSAB convened to formulate India's Draft Nuclear Doctrine. The draft doctrine was announced on 17 August 1999 and later a shorter version of the same was issued by the Cabinet Committee on Security in January 2003 along with the outline of its Command and Control structure four and a half years after the Pokhran-II series of tests in May 1998.

The two tenets of India's nuclear policy that have been widely publicised are the NFU policy and the credible minimum deterrent. Both were designed to project India as a responsible nuclear power.

Though the declaratory policy remains unchanged, the domestic Indian debate and technological developments suggest that the actual doctrine may not be in line with these principles anymore. Influential people like former defence minister Manohar Parrikar[5] have publicly questioned the wisdom of continuing with the NFU and have suggested that India should liberate itself from these self-imposed constraints over its nuclear use policy. Other former senior officials, such as former secretary external affairs and former national security advisor, Shiv Shankar Menon,[6] have also hinted at options such as a pre-emptive first strike by India. There is, therefore, a dichotomy between the declaratory doctrine and the actual operational doctrine but there appears to be no effort by India to clarify the prevailing uncertainty. The abandoning of the NFU policy could be viewed as 'unlearning' the politico-diplomatic considerations that informed the original policy. From another perspective, it could also be explained as a natural progression or adaptation of thinking in response to technological developments as well as changes in the security environment.

In the realm of command and control, India's learning appears to have been stymied by the continued jostling for control over the nuclear assets between the military and the civilian establishment and the perpetual inter-services rivalries. This has resulted in an anomalous situation where the strategic forces commander (a serving military officer) is directly answerable to the national security advisor (a civilian) rather than his military superiors. Sufficient information about the safety and security arrangements is also not available in the public domain and no specialised institutionalised structures are known to exist.

One area where learning is clearly discernible is strategic export controls. India has brought its strategic export controls regime in conformity with international norms and practices. It has traditionally been critical of the international non-proliferation regime for being discriminatory and has been castigating the technology control regime as a barrier to the acquisition of advanced technologies. However, it has learnt that instead of continuing to protest and criticise the inequities of the existing order, its interests would be better served by aligning its policies with the international regime. India's growing strategic affinity with the US and the political backing by the former has facilitated its entry into the international nuclear regime through the grant of an exceptional waiver by the NSG, thereby according it greater access to advanced technologies.

India is a state party to the Nuclear Safety Convention but it has not been able to implement its provisions effectively, especially with regard to establishment of an independent regulatory body for this purpose in accordance with the obligations accruing from being a party to the Convention. The convention specifically asks the states parties to take the appropriate steps to ensure, 'an effective separation between the functions of the regulatory body and those of any other body or organisation concerned with the promotion or utilisation of nuclear energy.'[7] However, India continues to rely on its Atomic Energy Regulatory Board, which is a subordinate organisation of the Atomic Energy Commission (AEC) and Department of Atomic Energy to perform the regulatory functions including licensing, inspections etc. The Manmohan Singh government had tabled a bill regarding the establishment of the Nuclear Safety Regulatory Authority in the lower house of the Indian parliament (Lok Sabha) in September 2011. However, over the following three years, the bill could not be passed and lapsed with the dissolution of the parliament in 2014. It is clear that the Indian government did not pursue the passage of this bill with the requisite vigour, most probably under pressure from a strong and influential nuclear establishment. The current BJP government has not shown any interest in reviving and pursuing the bill. This raises questions about the viability and integrity of India's regulatory regime. India could have learnt useful lessons from the Fukushima disaster in Japan which, among other reasons, could be attributed to lack of an independent nuclear regulatory body.

India has also been pursuing the development and acquisition of a BMD system since the late 1990s. Initially, it saw it as a means to developing closer affinity with the US by supporting the Bush administration's NMD programme and gaining access to high-end American military technologies. However, the bureaucratic and institutional interests of the DRDO have kept the programme alive; though, apparently unsatisfied by the progress of the domestic effort, India has recently signed a $5.5 billion deal with Russia for the purchase of five regiments of Russia's advanced S-400 air defence system. In pursuing the BMD systems, India seems to have ignored the lessons of the Cold War wherein the two super powers, having realised the destabilising potential of these systems, had mutually restrained their development and deployment through the instrument of the ABM Treaty. In the absence of any overarching strategic restraint mechanism in South Asia and lack of an arms control tradition, this Indian action is bound to spur a strategic arms race in the region. It would also create crisis instability, with each side itching to go first during a serious crisis. Had India drawn some inferences from the Cold War history, it might have refrained from introducing such destabilising systems of dubious value.

It appears that India's experience of habituating with the bomb over the last two decades has been a mixed one, with useful lessons learned in some areas—especially in its approach towards the non-proliferation regime—while limited learning appears to have taken place in others such as doctrine, command and control, safety and security, and regulatory regimes. In terms of doctrine, India has regressed from initial transparency to ambiguity and only limited information is available about the command and control organisation, protocols, and chain of succession. There is scant public knowledge about the existence of any purpose-built institution to oversee the safety and security, while the regulatory authority is constrained by lack of autonomy.

On another note, most of the learning in India, not unlike Pakistan, has been in the realm of 'simple learning', implying some adjustments in the means for the achievement of ends of policy. However, there seems to be lack of learning in the domain of 'complex learning' which involves drawing inferences from the experience of other nuclear powers as well as adjusting the ends in line with the demands of a nuclearised security environment. India does not seem to have learnt that in a mutual nuclear deterrence environment, there is no space for the use of military instruments to attain political objectives. However, frustrated by its inability to put its vast conventional advantage vis-à-vis Pakistan to any use, it has continued to seek space for a limited conventional war under the nuclear overhang, ignoring the fact that there is no guarantee that such a conflict will not escalate to a nuclear confrontation between the two South Asian antagonists. Instead, it is venturing into newer, more complex, and dangerous domains of nuclear competition by nuclearising the maritime domain. This clearly shows lack of 'complex' nuclear learning on India's part wherein it has been unable to make adjustments to the 'ends' of its policy along with modification of the means to be employed.

Notes and References

1. K. Subrahmanyam (ed.), *Nuclear Proliferation and International Security* (New Delhi: Lancer International, 1991); *Nuclear Myths and Realities: India's Dilemma* (New Delhi: ABC Publishing House, 1981); and *India and the Nuclear Challenge* (New Delhi: Lancer International, 1991).

2. K. Sundarji, 'Nuclear Weapons in Third World Context', *Combat Papers II and II* (Mhow, India: College of Combat, August 1981); K. Sundarji, *Blind Men of Hindoostan: Indo-Pak Nuclear War* (New Delhi: UBS Publishers Distributors Ltd, 1993).

3. Jasjit Singh, *Nuclear India* (South Asia Books, 1998); Jasjit Singh and Manpreet Sethi, *Nuclear Deterrence and Diplomacy* (Knowledge World, 2004).

4. Brigadier V. K. Nair, *Nuclear India* (New Delhi: Lancer International, 1992).

5. Special Correspondent, 'Why Bind Ourselves to "No First Use Policy", says Parrikar on India's Nuke

Doctrine', *The Hindu*, New Delhi, 10 November 2016.

6. Shiv Shankar Menon, *Choices: Inside the Making of India's Foreign Policy*, Series: Geopolitics in the 21st Century (Washington DC: Brookings Institution Press, 2016).

7. IAEA, INFCIRC/449, p. 4. Available at <https://www.iaea.org/sites/default/files/infcirc449.pdf>

Contributors

Naeem Ahmad Salik is currently a Senior Fellow at the Centre for International Strategic Studies (CISS), Islamabad. He completed his PhD from the University of Western Australia in May 2015. Between February 2009 and March 2011, he taught at the Department of Strategic and Nuclear Studies at the National Defence University (NDU), Islamabad. He has been a visiting scholar at the School of Advanced International Studies (SAIS) at Johns Hopkins University and a Guest Scholar at the Brookings Institution, Washington DC, from January 2006 to August 2008. He was also a Visiting Scholar at Stanford University in 2004 and at the Stimson Center in 1995. During the thirty-three years of military service, he gained experience of Command, Staff, teaching, and research positions. He was part of the team of officers that laid the foundations of Pakistan's Nuclear Command and Control system after the May 1998 nuclear tests. Subsequently, he served as Director Arms Control and Disarmament Affairs at the Strategic Plans Division until his retirement in October 2005. Brigadier Salik holds a Masters in History from the Punjab University (1981), a BSc Honours in War Studies from the Balochistan University (1985), and an MSc in International Politics and Strategic Studies from the University of Wales, Aberystwyth, UK (1989). He served as an Associate Professor at the Department of Defence and Strategic Studies, Quaid-i-Azam University, Islamabad, from 1994–96 and has since been teaching various courses related to strategy, nuclear policy, arms control, technological dimensions of warfare, and South Asian Security as a visiting faculty member. His areas of expertise include South Asian security with special reference to Indian and Pakistani policies on nuclear and missile issues, nuclear and conventional confidence building measures, arms control and disarmament issues, nuclear non-proliferation regimes, nuclear safety and security, and export controls. His books *Genesis of South Asian Nuclear Deterrence—Pakistan's Perspective* (2009) and *Learning to Live with the Bomb* (2017) have been published by Oxford University Pakistan. He has published over thirty research articles in reputed national and international publications. His most recent book—an edited volume titled *Nuclear Pakistan—Seeking Security and Stability* (University of Lahore Press) was released in March 2018.

Sitakanta Mishra is currently a Professor of International Relations at the School of Liberal Studies (SLS), Pandit Deendayal Petroleum University (PDPU), Gandhinagar, Gujarat (India), and Managing Editor of the Liberal Studies Journal. Previously, he was a Research Fellow at the Centre for Air Power Studies (CAPS), New Delhi, and guest faculty at the Nelson Mandela Centre for Peace and Conflict Resolution, Jamia Millia Islamia, New Delhi. He was a Visiting Research Scholar at the Cooperative Monitoring Centre, Sandia National Laboratories USA, during September-December 2013, and CRDF (USA) Visiting Scholar during April-August 2014. He is also the South Asia liaison of the International Network of Emerging Nuclear Specialists (INENS), London. Dr Mishra received his PhD from the Centre for South Asian Studies of Jawaharlal Nehru University (JNU), New Delhi, in 2011. He is an Associate Member of the Pugwash India, and Non-Resident Research Fellow at the Centre for Air Power Studies, New Delhi. Formerly, he was also the Associate Editor of the Indian Foreign Affairs Journal published by the Association of Indian Diplomats (AID), Sapru House, New Delhi. Dr Mishra has authored three books, *Defence Beyond Design: Contours of Nuclear Safety and Security in India* (2017); *Parmanu Politics: Indian Political Parties and Nuclear Weapons* (2015); *Cruise Missiles. Evolution, Proliferation, and Future* (2011); and *The Challenge of Nuclear Terror* (2008). He is also authored many national and international research papers related to nuclear safety-security, terrorism, nuclear energy, nuclear proliferation, cruise missiles, Indian foreign policy, and South Asian affairs. He is also involved in Indo-Pak Track-II dialogue.

Ali Ahmed has been an infantry officer, an academic, and an international civil servant. He blogs at www.ali-writings.blogspot. in on issues in strategic and peace studies. He is the author of *India's Doctrine Puzzle: Limiting War in South Asia* (Routledge 2014). He has a doctorate in international politics from the Jawaharlal Nehru University, New Delhi.

Zafar Ali has been working at various positions in the government. He holds a Master's degree and PhD in Defence and Strategic Studies from Quaid-i-Azam University, Islamabad. He has participated in various bilateral and multilateral forums including national/international workshops and seminars on non-proliferation, security, and strategic export control related issues. He is a former visiting fellow of the Henry L. Stimson Center, Washington DC, the Center for Non-Proliferation Studies, James Martin Center, Monterey, California, and former visiting fellow of the Center for International Trade Security, Georgia, USA. He has been on the visiting faculty of the National Defence University and Quaid-i-Azam University, Islamabad.

Happymon Jacob is Associate Professor of Diplomacy and Disarmament Studies at the Jawaharlal Nehru University (JNU), New Delhi. Prior to joining JNU in 2008, Jacob held teaching positions at the University of Jammu in J&K and the Jamia Millia Islamia University, New Delhi; and research positions at the Centre for Air Power Studies, Delhi Policy Group and Observer Research Foundation—all based in New Delhi. Jacob serves as a member of the international Governing Council of the Nobel Peace Prize-winning Pugwash Conferences of Science and World Affairs. He is the Honorary Director of the India chapter of the Chaophraya Dialogue—an India–Pakistan Track II Initiative which has been running since 2008. Jacob curates the recently launched Indo-Pak

Conflict Monitor—an independent research initiative to monitor ceasefire violations, conflict patterns, and escalation dynamics between India and Pakistan. He is also a columnist with *The Hindu*, one of India's leading English-language dailies. His book, *Line on Fire: Ceasefire Violations and India-Pakistan Escalation Dynamics* was published in March 2019 by Oxford University Press India.

Tanvi Kulkarni is a PhD candidate in the Diplomacy and Disarmament Division, Centre for International Politics, Organisation, and Disarmament of the School of International Studies, Jawaharlal Nehru University (JNU) in New Delhi. Her doctoral thesis 'Assessing the Impact of Crises and International Norms on Nuclear Confidence Building Measures' examines the prospects of NCBMs between nuclear dyads like India and Pakistan. In August 2014, Tanvi completed her Masters of Philosophy (Diplomacy and Disarmament) from JNU and submitted her dissertation on a study of credible minimum deterrence and no first use in India's nuclear doctrine. Her research interests concern South Asia's nuclear politics, India's nuclear weapons policy and doctrine, and India–Pakistan nuclear confidence building measures. Tanvi has previously worked with the Institute of Peace and Conflict Studies, New Delhi; she was a visiting fellow at the Stimson Center, Washington DC in March 2015 and contributes commentaries on the online portal 'South Asian Voices' curated by the Stimson Center. She is a consultant coordinator for the India chapter of the Chaophraya Dialogue—an India–Pakistan Track II initiative, and a project coordinator for the Indo-Pak Conflict Monitor.

Index

A

All Spectrum Missile Defence 101
American Consulate 22, 55
APSARA 21, 143
Arsenal size 63
Atomic Energy Act (1962) 30, 112, 114, 116, 119, 128, 144, 146, 151, 154, 164

B

Babri Masjid 68
Ballistic Missile Defence (BMD) 9, 11, 28, 101, 107, 167, 171, 177, 184
Bhabha, Homi 18, 20–2, 32, 38–9, 42, 49–50, 58, 93, 143
Bhabha Atomic Research Centre 42, 93, 97, 144, 149–52, 165
Bharatiya Janata Party (BJP) 36, 44–5, 47–8, 57, 64, 68, 70–1, 106, 169, 188
Brahmos 121, 128
Breeder reactor technology 20

C

Cabinet Committee on Security (CCS) 59, 62, 83, 94–5, 119, 186
Cannisterised 96
Central Intelligence Agency (CIA) 22–3, 39, 44, 47, 52, 55–8
Chemical Weapons Convention 116, 118

Chiefs of Staff Committee (CoSC) 10, 71, 75, 94, 182
Chinese bombers 28
Chinese nuclear capability 25, 49
CIRUS 21, 23–6, 30, 34, 36, 141, 143
Civil–military relations 73, 86–7, 106
Civil Nuclear Agreement 30, 52, 54
Cold Start 60, 63–4, 66, 68, 70, 76–8, 80, 82–3
Command and Control Structure 1, 3, 14, 52, 92, 99, 186
Confidence Building Measures (CBMs) 89
Counter insurgency 64, 76, 78
Counter retaliation 69
Credible minimum deterrent 62, 186
Cruise Missile Defence (CMD) 102
Customs Act (1962) 114

D

De-escalation 61, 82
Defence budget 70
Defence Research and Development Organisation (DRDO) 51, 93, 95–7, 101, 105–7, 119, 174, 176, 183, 189
Defense Intelligence Agency (DIA) 43, 57, 103
Defensive systems 28, 168, 177
Delivery system 19, 25, 27–8, 36, 40, 72, 95–6, 112, 118, 172
Department of Atomic Energy (DAE) 20, 27, 114, 116–8, 131, 160, 163

Deterrence 3, 5, 12, 59–61, 63–4, 67, 69, 74, 79, 81, 83, 85, 89–90, 99–100, 106, 179–82, 190
Doctrinal thinking 9, 75–6, 79
Draft nuclear doctrine 1, 14, 52, 69, 74, 92, 186

E

Early warning system 28
Enhanced nuclear detonation safety system 97
Escalatory control 61
European markets 26
Export controls 9, 11, 15, 41, 111, 113, 115–8, 124, 126, 128–9, 187

F

Fast breeder reactors 53, 123, 127, 131, 150
Fissile material 21, 36, 41, 53, 90–1, 123, 139, 147
Foreign Trade (Development and Regulation) Act–1992 (FTDR Act–1992) 113–4
Fuel supplies 43
Full spectrum deterrence 69, 182

G

Gandhi, Prime Minister Rajiv 5, 43, 51, 76, 139, 161
Global powers 8
Gross domestic product (GDP) 70

H

Hiroshima and Nagasaki 19
Hydrogen Bomb 45, 72–3

I

Import and Export Act (1947) 113
Indian Department of Atomic Energy 27
Indian Nuclear Energy Programme 22–3, 151
India's Nuclear Bomb 5, 13, 58, 84
India's strategic export control regime 110
Intelligence Coordination Group (ICG) 103
International Atomic Energy Agency (IAEA) 41, 53, 116, 123, 127–8, 132–40, 148, 155, 158–63, 191
International learning 2
International Panel on Fissile Materials (IPFM) 53, 90

J

Jacob, Happymon 11, 86–7, 130, 164, 166
Jammu and Kashmir 7, 46, 66, 78, 173

K

Kargil 7, 14, 66, 77, 79
Kulkarni, Tanvi 11, 130

L

Learning to Live with the Bomb 8, 14
Line of Control (LoC) 66

M

Madras University 5
Massive retaliation 9, 74
Military nuclear programme 22, 24, 26–8, 36, 126, 151
Military standoff 7

Ministry of External Affairs (MEA)
    93–5, 119
Mishra, Dr Sitakanta 10, 89, 166
Missile developments 28
Mixed Oxide Fuel (MOX) 43
Mumbai crisis 7
Mutually assured destruction (MAD)
    66, 70, 82

N

National Defence College 49, 74
National Democratic Alliance 99
National Democratic Alliance (NDA) 61
National Disaster Response Force 69
National Security Advisor (NSA) 1, 10,
    67, 71, 73, 94–5, 168, 174, 187
National Security Advisory Board
    (NSAB) 6, 62, 66, 95
National Security Council 26, 44–5, 71,
    94–5
National Security Council Secretariat
    (NSCS) 71, 95
Negative security assurance 62
No first use (NFU) 1, 4–6, 62, 65, 71,
    74, 79, 81–2, 90, 93, 174, 179,
    186–7
Non-proliferation regime 50, 90, 123,
    140, 187
Non-Proliferation Treaty 113
NSG waiver 141, 162
Nuclear: abstention 24, 49; ambitions
    9, 40, 53; -armed submarine 70,
    180; Command Authority 94–5,
    107; delivery systems 19, 36, 40;
    deterrence 5, 67, 69, 89–90, 99,
    179, 181, 190; detonation 16, 29, 97
    Energy Commission 18; escalation
    69; Export Control Regime 10;
    policy 3–4, 8, 44, 54, 62, 85,
88, 179, 186; powers 7, 18, 32;
    regulatory regime 9–11, 130, 134,
    137–9, 142, 152; security summits
    91, 135; strategy 4, 6, 75, 83, 86, 95,
    100, 106; weapons 2, 4–5, 7–8, 10,
    12, 14–19, 22–6, 31–2, 36, 39–44,
    49–52, 55–6, 59, 62–3, 65, 68, 70–
    2, 74, 78–9, 81, 83, 86–7, 89–94,
    96–8, 101, 104–7, 123, 139–40,
    161, 172, 190
Nuclearisation 1, 3, 5, 7, 48–9, 73, 79,
    182, 186
Nuclear learning 2–4, 7–9, 12, 14, 16–
    17, 60, 190; in India 4, 12; in South
    Asia 3, 7

O

Operation Parakram 65, 69
Organisational interests 60

P

Peaceful nuclear explosion (PNE) 1, 22,
    28, 30–2, 35–6, 49
Perkovich, George 5–6, 13, 50, 58, 84
Permissive Action Link (PAL) 97
Plutonium 21–6, 30, 33–4, 36, 40, 53,
    55, 58, 90–1, 105, 163; reprocessing
    plant 24; separation plant 24;
    Separation in Nuclear Power
    Programs 53
Pokhran–II 57, 186
Pokhran test 40
Political: leadership 4, 10, 92, 175, 180;
    system 92, 109
Pressurised heavy-water reactors
    (PHWR) 53, 131, 144, 146, 150
Procurement 26, 122

R

Rajya Sabha 25
Rao, Narasimha 6, 40, 43, 51, 93
Rashtriya Rifles and Assam Rifles 69
Reactor-grade plutonium 53, 91
Regional power 69
Research and Analyses Wing (R&AW)
    103
Research reactor 21, 143

S

Safety and security 2–3, 10, 15, 52, 93–
    4, 96–7, 132, 134–6, 145, 155–6,
    158, 187, 189
Security environment 1, 8–9, 14–15, 44,
    91, 184, 187, 190
Senate Intelligence Committee 43
Singh, Jasjit 6, 58, 79, 83, 108, 186, 190
Singh, Prime Minister V. P. 5
Sir Dorabji Tata Trust 20
Sketchy Passive Defence 100
Smiling Buddha 18
Special national intelligence estimate
    (SNIE) 24
State Department 23, 25–6, 28–30, 47
Strategic Armament Safety Authority
    94–5
Strategic command, control, and
    communications 19
Strategic Force Command 10

Strategic nuclear plants 93
Strategic planning staff (SPS) 94–5
Strategic trade control system 109
Subrahmanyam 4–6, 32, 35–6, 49, 54,
    56, 58, 73–4, 87, 170, 186, 190
Sundarji, General K. 4–5, 12, 77, 85,
    87, 186, 190
Surgical strikes 66, 77

T

Tactical nuclear weapons (TNW) 59
Targeting philosophy 6
Tibetan plateau 70
Trinity 19

U

United Alliance 61
Unsafeguarded: power reactors 25;
    reactor 42; uranium 24
Uranium fuel 21, 24, 33, 43

W

Warhead 62, 98
Wassenaar Arrangement 52, 110, 112,
    119, 121, 128, 141
Weaponisation 51
Weapons development 24, 55
Weapons-grade plutonium 21, 24